Ecology of Wildfire Residuals
in Boreal Forests

Ecology of Wildfire Residuals in Boreal Forests

Ajith H. Perera and Lisa J. Buse

Ontario Forest Research Institute
Sault Ste. Marie, Ontario
Canada

WILEY Blackwell

Library of Congress Cataloging-in-Publication Data

Perera, Ajith H.
  Ecology of wildfire residuals in boreal forests / Ajith H. Perera and Lisa J. Buse.
      pages cm
  Includes bibliographical references and index.
  ISBN 978-1-4443-3653-5 (cloth)
1. Fire ecology. 2. Taiga ecology. I. Buse, Lisa J. II. Title.
  QH545.F5P47 2014
  577.2'4–dc23
                                                  2013050116

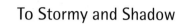
To Stormy and Shadow

Contents

Acknowledgments

We have many to thank for their help in writing this book.

First we thank Alan Crowden who was the prime motivator; without our regular chats during IALE meetings, this book would not have materialized.

Several colleagues helped us prepare the manuscript: Rob Routledge (literature search), Marc Ouellette (maps and technical support), Trudy Vaittinen (graphics support), and Geoff Hart (editorial support). We thank them all for their most valuable contributions. Also, we thank the following for permitting us to use their photographs: Jeff Bowman, Mike Francis, Dave Ingram, Stephen Luk, David Maddison, Antoine Nappi, Bill Parker, Gerry Racey, Rob Routledge, Dave Schroeder, and Peter Uhlig.

We most gratefully acknowledge the many colleagues who reviewed parts of the manuscript, and provided us with thoughts to improve its content. They are: Jim Baker, Ken Baldwin, Den Boychuk, Herb Cerezke, Joe Churcher, Dave Euler, Rob MacKereth, Doug McRae, Antoine Nappi, Brian Naylor, Bill Parker, Sunil Ranasinghe, Michel Saint-Germain, and Michael Ter-Mikaelian. In addition, we benefitted much from discussions with Don Bazeley, Joe Churcher, Doug McRae, Antoine Nappi, and Ken Raffa.

We are indebted to our friend Susan Smith, who volunteered to read the whole book manuscript and offered many suggestions to improve its readability and clarity.

Finally, we thank Wiley-Blackwell staff, especially Kelvin Matthews, for assistance in publishing the book.

About the companion website

This book is accompanied by a companion website:

www.wiley.com/go/perera/wildfire

The website includes:
- Powerpoints of all figures from the book for downloading
- PDFs of tables from the book

1 Introduction

Wildfires and their role in influencing ecological processes and patterns in the boreal forest biome have drawn the attention of scientists for several decades. Much has been written on this topic, including texts by Wein and McLean (1983a), Johnson (1992), Goldammer and Furyaev (1996a), and Kasischke and Stocks (2000a), to address the physical processes that occur during and after wildfires, the effects of wildfires on forest ecosystems, the post-fire recovery process, and even the management of wildfires to minimize their economic impacts. Such focus on the physical process of wildfire and its ecological impacts may create a general impression that boreal wildfires destroy all components of boreal forest ecosystems. This is not true in most instances; many components of the forest vegetation, both above and below ground, remain alive. Many plants either escape or survive the fires and most vegetation, even if burned, retains some or all of its pre-burn structure.

The vegetation that is not burned, together with the biomass that is not completely incinerated, creates fine-scale heterogeneity within the areas burned by boreal wildfires. These post-fire remnants are a critical component of the local ecosystem recovery processes after wildfire, and contribute significantly to several ecological processes in the broader landscape beyond the spatial signature of a fire. Recently, boreal wildfire residuals have attracted increasing scientific,

Ecology of Wildfire Residuals in Boreal Forests, First Edition. Ajith H. Perera and Lisa J. Buse.
© 2014 Ajith H. Perera and Lisa J. Buse. Published 2014 by John Wiley & Sons, Ltd.
Companion Website: www.wiley.com/go/perera/wildfire

social, and even political attention, from the perspective of conservation biology, because of attempts to emulate post-fire patterns during forest harvesting and concerns about the ecological effects of harvesting the residuals. However, neither the presence of such post-fire residuals nor their ecology has been well recognized in the fire literature; most synthetic reviews of studies of boreal wildfires have focused on the fire process or the ecological effects of fire, not on what is left unburned.

In this context, our focus in this book is the patterns and ecological processes related to the residuals created by boreal wildfires, and the use of this knowledge in forest management. Here we explore what is known and, in so doing, identify the many questions that remain about the ecology of wildfire residuals in the boreal forest biome. We use the term "wildfire residuals" to signify all of the vegetation structure remaining after a fire: still living, in the process of dying, and killed by the fire.

In this first chapter, we briefly introduce two topics that are key to understanding boreal wildfire residuals, especially for those unfamiliar with this forest type: the boreal forest biome – its occurrence, distribution, and general characteristics; and boreal wildfire – its general characteristics and ecological significance. We then provide an overview of the goals, scope, and structure of the book.

The boreal forest biome

Though not as rich in biota as other terrestrial biomes, boreal forests still harbor many plant and animal species. Boreal forests are among the least populated of all forest biomes. Considerable portions remain uninhabited and natural ecological processes continue to occur largely without human interference, especially in North America and Russia (UNECE 2000). These forests have attracted wide interest from scientists, land managers, politicians, and the general public. Not only are boreal forests an important source of commodities ("natural resources") such as timber, oil, and minerals, they are home to many communities of both native (indigenous) and immigrant populations, the latter drawn mainly by industries based in natural resource extraction.

More recently, boreal forests have become valued in relation to such global phenomena as biodiversity and climate change. These forests are thus considered important from a wide range of perspectives: ecological, for example for wildlife and plant habitat, and carbon budgets; economic, for example for forestry, agriculture, mining, and tourism; and social, including for settlement and recreation. Furthermore, boreal forests are also of political interest with respect to such issues as the rights of indigenous peoples, the debate over the conservation of

INTRODUCTION

biodiversity versus exploitation of natural resources, and the global role of these forests in climate change.

The boreal forest biome is one of the largest pools of organic carbon on Earth and is, therefore, a significant factor in regulating global and regional climates. Estimates indicate that the carbon stored in the boreal biome totals over 700 Pg, of which about 82 Pg is stored in above- and belowground plant biomass, about 200 Pg in soil, and about 420 Pg in peatlands. Carbon stored in the boreal biome thus accounts for over one-third of the total terrestrial global carbon pool (Apps *et al.* 1993). The forested portion of the boreal biome alone contains almost 300 Pg, which is equivalent to nearly 16% of the world's total (Kasischke *et al.* 1995).

In addition to this immense carbon pool, the boreal biome supplies many other ecosystem goods and services. It acts as a reservoir for biological and genetic diversity, purifies the air and water, and provides habitat for wildlife, including birds, mammals, insects, and microorganisms. The boreal biome also provides food and renewable raw materials for human use; it is a major source of softwood timber and provides jobs and economic stability to many rural and remote communities. For indigenous peoples, the boreal forest is a source of livelihood as well as the base for their culture and spiritual sustenance.

The boreal biome is unique in that it remains relatively unpopulated. Throughout most of the biome, there is more than 2 ha of forest per inhabitant (UNECE 2000). In addition, it is largely inaccessible: only about one-third of the world's boreal forests are within 10 km of any transportation infrastructure (FAO 2001), and this contributes to the fact that much of the forest remains relatively undisturbed by humans. In fact, the overwhelming majority of the global temperate and boreal forests that are "undisturbed by man" are located within Russia (81%) and Canada (13%) (UNECE 2000). Also, most of the boreal forest is under public ownership: in Canada and Russia, >90% of the forest is publically owned. The exception is the Nordic countries (e.g., Finland, Norway, and Sweden) where ≥60% is privately owned (UNECE 2000).

According to recent figures, the forested area within the boreal biome remains relatively stable, changing by ±0.5% annually (UNECE 2000). However, the boreal forest is slowly becoming more accessible and, as a result, its use is increasing. Historically, those who colonized the area survived by hunting, fishing, undertaking shifting cultivation of a few crops, and raising cattle in forest pastures, all activities that caused little change to the boreal forest environment, although evidence exists that fire was used to clear land and improve pastures (Kuusela 1992). By the 19th century, the forests were being cleared more extensively to permit agriculture, to pasture livestock, and to extract wood for house construction and fuel and to provide wood products to construct ships. During the 20th century, increasing industrialization and the growing demand for raw materials resulted in increased

INTRODUCTION

extraction of wood from boreal forests (Kuusela 1992). This demand continues to escalate, resulting in increasing pressure to expand the use and management of boreal forests and also to protect forest resources from natural disturbances, primarily wildfires (UNECE 2000).

Geographical distribution

The boreal biome is one of three major global forest types, together with the tropical and temperate forests, and accounts for one-third of the world's forest area (FAO 2001) and about three-quarters of all coniferous forests (Kuusela 1992). In North America, this biome is generally referred to as the *boreal zone*, whereas in Eurasia it is commonly named the *taiga*, a Russian word that was first associated with the coniferous forests of Siberia (Shorohova *et al.* 2009) and that is said to mean "vast, all-over forests, impassable to men" (Hytteborn *et al.* 2005). This biome stretches in broad transcontinental bands across North America and Eurasia (Figure 1.1). In North America, the boreal biome extends throughout eastern and central Canada, and in western Canada, it extends northward into Alaska. In Eurasia, the biome extends throughout Fennoscandia (Norway, Finland, and Sweden) and Russia, as well as into the northern parts of China, Kazakhstan, and Mongolia (Brandt 2009).

Some authors describe the bounds of the boreal biome in general terms, for example, the forest that occurs south of the tundra and north of the deciduous forests and grasslands. Others cite more specific latitudinal bounds of between about 45°N and 65°N, but with considerable regional variation. In North America close to half of the boreal zone occurs between 45°N and 55°N, whereas in Eurasia most occurs north of 55°N (Goldammer and Furyaev 1996b). Still other authors suggest that the northern and southern boundaries of the boreal forest are determined by long-term average values of various climatic factors, most notably summer temperatures. For example, Larsen (1980) indicated that the northern boundary of the boreal zone corresponds to the July 13 °C isotherm, whereas the southern boundary coincides approximately with the July 18 °C isotherm. In North America, the northern boundary has also been associated with the modal July position of the front that separates the continental Arctic and maritime Pacific air masses, whereas the southern boundary is associated with the winter position of the Arctic front (Bonan and Shugart 1989, citing Bryson 1966). Similar climatic relationships have been found for the Eurasian boreal forests (Krebs and Barry 1970).

Given the different approaches used to describe the location of the boreal biome, it is not surprising that estimates of its total extent vary from 10.0 million to 14.7 million km² (Table 1.1). The discrepancies in these estimates can be

Figure 1.1 The boreal biome in the northern hemisphere occurs in broad bands across northern North America and Eurasia. (NRDC 2004. Reproduced with permission of the Natural Resources Defense Council.)

Table 1.1 Examples of published estimates of the global extent of boreal forest (taiga).

Area included	Estimate of extent ($\times 10^6$ km^2)	Source
Northern boreal forests	10.0	UNECE (2000)
Forested area of boreal biome	11.7	Kasischke *et al.* (1995)
Boreal forest	12.0	Goldammer and Furyaev (1996b)
Boreal forest	13.0	FAO (2001)
Boreal biome	14.3	Kasischke *et al.* (1995)
Boreal forest	14.7	Bonan and Shugart (1989)
Boreal forest	14.7	Bourgeau-Chavez *et al.* (2000)

INTRODUCTION

attributed in large part to varying definitions of what constitutes the boreal biome: some include non-forested vegetation types such as peatlands; others include the hemiboreal region, which is the ecotone between the boreal and the nemoral (broad-leaved deciduous) forests (Hytteborn *et al.* 2005); and some exclude China and Mongolia (e.g., Kuusela 1992, UNECE 2000).

Much of the global boreal forest cover is thought to occur in Russia, followed, in descending order, by North America, northeastern China, and Scandinavia (Denmark, Norway, and Sweden) (Goldammer and Furyaev 1996b). The Russian boreal forest area is estimated at 9 million km^2 (Shorohova *et al.* 2009). In North America, a very detailed analysis of the extent of the boreal zone indicated that it covers 6.3 million km^2, over half of which is forest and other wooded land (Brandt 2009). A less detailed estimate indicated that the total area of boreal forest in Scandinavia is just over 0.5 million km^2 (Esseen *et al.* 1997).

The boreal forest biome is often described as having three distinct zones from south to north: the closed-canopy forests in the south, the more open lichen woodlands in the central region, and the sparsely treed forested tundra in the north (Larsen 1980). Brandt (2009) distinguished the *boreal zone*, which represents the broad circumpolar vegetation zone at high northern latitudes, from the *boreal forests* (plural), which represent the forested areas within the boreal zone. Although mainly covered with trees, the boreal zone also includes large areas of lakes, rivers, and wetlands, as well as naturally treeless terrain such as alpine areas and peatlands, with grasslands in drier areas. Brandt considered the term *boreal forest* (singular) colloquial and confusing, since it has been used to refer both to the forested area within the boreal zone and to the boreal zone itself. Nonetheless, in this book we use the term "boreal forest" to refer to the treed areas of the boreal biome and have excluded the sparsely treed tundra.

Distinguishing features

Geoclimate

The boreal forest biome is characterized by long, cold winters and short, moderately warm summers (Larsen 1980). Temperatures fluctuate both seasonally and annually. These fluctuations are most pronounced in the continental climates of interior Alaska and eastern Siberia, where seasonal temperature extremes range as much as 100 °C, and less extreme in Scandinavia and eastern Canada, where a more oceanic climate prevails (Bonan and Shugart 1989, citing Rumney 1968). Average July temperatures in the taiga range from 10 to 20 °C (Hytteborn *et al.* 2005), and average annual temperatures in the boreal forest globally range from just below to just above freezing, with much regional and local variation. Although the growing season is short, the days during that period are long, in some areas exceeding 16 h.

INTRODUCTION

Annual rainfall in the boreal biome is relatively low, and more than half the annual precipitation falls during the summer (Bonan and Shugart 1989, citing Rumney 1968). In North America, precipitation decreases from east to west: in eastern Canada, annual averages range from 510 to 890 mm; in central Canada from 380 to 510 mm; and in western Canada from 180 to 380 mm. In Eurasia, the pattern is reversed, with annual rainfall of more than 510 mm occurring west of the Ural Mountains, slightly less east of the Urals, and the least (from 130 to 250 mm) in northeastern Siberia (Bonan and Shugart 1989, citing Rumney 1968). The long, cold winters allow snow to accumulate and, as in the Russian taiga, snow cover can remain for up to 240 days (Hytteborn *et al.* 2005).

The boreal biome is geologically young and is still changing. In western Canada and Alaska, it developed during the Holocene (i.e., within the last 12 000 years), and in eastern Canada, between 4000 and 8000 years ago, following deglaciation (Payette 1992). The same is true for Eurasia, where much of the boreal forest has developed since the last ice sheets receded (Hytteborn *et al.* 2005). The terrain is predominately flat to gently rolling, with the exception of four mountain chains: the Ural Mountains in Russia, a series of mountains in eastern and southern Siberia, the northern Rockies in western Canada, and the Scandes Mountains in Fennoscandia.

Where there is mineral soil, it is typically shallow, and alternates with wetlands and poorly drained organic soils. Soils are generally acidic and cold. Soil formation is characterized by *podzolization*, which is the movement of dissolved organic matter and minerals from the surface into the deeper soil horizons (Shorohova *et al.* 2009). Lignin in fallen conifer needles causes them to decompose slowly, creating a mat over the soil. Tannins and other acids cause the upper soil layers to become very acidic, and permanent shade from the canopy slows evaporation, keeping the soils wet. The combination of cold and wet conditions slows the decomposition of organic matter and limits nutrient cycling (van Cleve and Dyrness 1983, Shorohova *et al.* 2009). Over time, the organic layers slowly accumulate and store huge amounts of carbon. In the northernmost areas of the biome, permafrost is prevalent, but trees can grow wherever the soil thaws to a depth of 1 m during the growing season (Shorohova *et al.* 2009).

Species composition and vegetation

The boreal forest biome includes coniferous forests and open woodlands interspersed with abundant wetlands and lakes. Relatively few tree species occur in the boreal forest. These are primarily conifers that are adapted to the cold, harsh climate and the thin, acidic soils. The main tree species are pines (*Pinus* spp.), spruces (*Picea* spp.), larches (*Larix* spp.), and firs (*Abies* spp.), which are sometimes mixed, usually after a disturbance, with deciduous hardwoods such as birches (*Betula*

INTRODUCTION

spp.), poplars (*Populus* spp.), willows (*Salix* spp.), and alders (*Alnus* spp.) (Larsen 1980, Goldammer and Furyaev 1996b).

Throughout the boreal biome, pines dominate on the dry sites and spruces in the wetter areas. In Russia, the forests are classified into "dark" or "closed" coniferous forests and "light" or "open" coniferous forests. The closed forests are spruce-dominated in western Europe and mainly comprise spruce–fir–pine mixed forests in Eastern Europe and Siberia. The open forests are pine- and larch-dominated (Hytteborn *et al.* 2005, Shorohova *et al.* 2009). In North America, spruce dominates in Alaska and central Canada. Fir and larch occur only in central and eastern Canada, and pine occurs everywhere except Alaska. Deciduous species, mainly aspen and birch, occur throughout the forest, typically invading newly disturbed areas (Bourgeau-Chavez *et al.* 2000). In Fennoscandia, the main species are Scots pine (*Pinus sylvestris*) and Norway spruce (*Picea abies*), but in the south, a combination of scattered occurrences of various broad-leaved trees occur among the conifers. In both the middle and northern boreal zones of Fennoscandia, the conifers dominate, with birch as the main broad-leaved species. These two zones are differentiated not by tree cover but by the understory plants, with the northern zone containing more willow thickets and various tall-herb communities and lacking the more southerly low-herb spruce and blueberry (*Vaccinium* spp.) forests (Esseen *et al.* 1997).

Throughout the boreal forest, mosses and lichens commonly cover the ground under the tree canopies, where they bind nutrients and insulate the soil. Low soil temperatures limit both water and nutrient uptake (Bonan and Shugart 1989) and, in combination with the short growing season, result in very slow tree growth. The estimated net annual increment of boreal forests worldwide is less than $2\,m^3\,ha^{-1}$, with slightly higher growth rates of $2-6\,m^3\,ha^{-1}$ reported for Scandinavian boreal forests (UNECE 2000). The estimated average growing stock (the sum of the volumes of all living trees) in the boreal forest ranges from 100 to $150\,m^3\,ha^{-1}$, except in Finland, where it is lower at 50 to $100\,m^3\,ha^{-1}$ (UNECE 2000). Despite these slow growth rates, boreal forests supply major quantities of wood and are valued for many other ecosystem services.

Compared with the forested ecosystems of southern latitudes, the patterns of species composition in boreal forests are relatively homogenous and simple. The heterogeneity in boreal forest vegetation communities results primarily from variability in climate, geology, and topography. Broadly defined geographically by latitude and climate, the boreal forest biome is frequently subject to extensive allogenic disturbances, such as windstorms, insect epidemics, and wildfires. These change the vegetation pattern over time, creating a "shifting spatial mosaic" of tree species assemblages and ages as a result of the periodic disturbances. The ensuing broad-scale spatial and temporal heterogeneity of vegetation patterns, and thus of ecological processes, induced by disturbances is a defining characteristic of the boreal forest biome (Bonan and Shugart 1989).

Boreal wildfires

Microclimate, site conditions, and local species composition vary immensely within this vast biome, and many regional and sub-regional differences exist in landscape patterns and ecological processes. However, wildfires have long been recognized as an integral part of ecosystems throughout the boreal forest biome; the scientific literature is rich with historical articles and books on boreal wildfires in North America and Fennoscandia. Although there must be an equally rich literature from Eurasia, those publications have not been readily available to readers in the western hemisphere. It is only in the recent past that detailed information on boreal forest fires has been available from Russia, and knowledge of wildfires in boreal China continues to be scarce in the English-language literature.

Despite the regional differences, it is generally accepted that boreal forests are predisposed to wildfire for two reasons (Scotter 1972, Wein and McLean 1983b, Johnson 1992, Goldammer and Furyaev 1996b). First, the long days of summer subject boreal forests (north of 50°N) to more than 15 h of insolation, which rapidly dries the fuel beds and lowers their moisture content, rendering them highly combustible. In addition, storms accompanied by lightning are common from early spring to early fall, providing a steady supply of ignition sources when the fuels are dry. Secondly, as the home for more than two-thirds of the word's coniferous forests, boreal forests are naturally rich in highly flammable fuel. For example, conifer needles containing waxes and terpenes cover the ground, in the form of loosely packed slow-decomposing litter, and the tree crowns have low-hanging branches. Consequently, wildfires are ubiquitous in boreal forests around the world.

Major characteristics

The types of boreal wildfires are also believed to exhibit similar trends among regions. As Wein and McLean (1983b) noted in their review, spruce-dominated forests in both boreal North America and boreal Europe usually support high-intensity crown fires. On the other hand, pine- and larch-dominated boreal forests on both continents typically do not support crown fires due to the high

INTRODUCTION

degree of self-pruning of the pines and the high moisture content and low resin content in larches. Typically, crown fires are more common in boreal North America and western Russia, whereas ground fires are more common in eastern Russia and Eurasia. Fennoscandian wildfires are mostly ground fires.

Recent information (de Groot *et al.* 2013) suggests that differences in boreal fire regimes between continents extend to the seasonality, fire sizes, fire intensity, and fuel consumption. The frequency of fires is generally high in pine-dominated forests, with the period between fires at a given location as short as 20–60 years. Spruce-dominated forests have a lower fire frequency, with longer periods between fires at a given location, and an estimated interval of about 100 years between fires. In areas dominated by wetlands and mixed deciduous–coniferous forest, the fire frequency can be even lower. There may be forest areas with a very low frequency of fires based on historical records, but these are the exception rather than the rule in the boreal forest biome. Except in the areas that are heavily inhabited by people, most boreal wildfires originate from lightning in Europe (Gromtsev 2002) and North America (Stocks *et al.* 2003).

The extent of boreal wildfires is vast, totaling many millions of hectares per year globally. Kasischke (2000) noted two other general trends in the extent of boreal wildfires using North American examples. First, the total area burned in relatively dry years is much larger than that in years with average precipitation. Secondly, the majority of the total area burned is accounted for by a few large fire events; single fires covering 100 000 ha are common in the boreal forest biome, and those that cover 1 million ha may occur during severe fire years. However, a reader searching for a simple answer to "how much of the boreal forest is burned by wildfires?" is likely to be confused. This is because global estimates of the wildfire area vary widely depending on the period, the methods for recording and determining the area of the fires, and the terminology used (Conard and Ivanova 1997, Shivedenko and Nilsson 2000). Goldammer and Furyaev (1996b) concluded that the largest extent of boreal wildfires occur in North America and Eurasia (mostly in Russia), with only a negligible extent in Fennoscandia and boreal China. Table 1.2 summarizes the boreal wildfire estimates reported in the scientific literature during the last three decades.

Much of the variability in past estimates of the extent of boreal wildfires stems from the unreliability of the data sources, namely the government agencies that track and record fire information (Conard and Ivanova 1997, Alexeyev *et al.* 2000). However, this problem should diminish over time. Rapid advances in remote-sensing technology and the increasing accessibility of data and decreasing cost of access to these data will increasingly let researchers independently monitor and collect information about boreal wildfires (Ahern *et al.* 2000, Sukhinin

Table 1.2 Estimates of the annual area burned by wildfires in the boreal forest biome, by region.

Boreal forest biome region	Wildfire area estimate ($\times 10^6$ ha year^{-1})	Period	Source
Pan-boreal	1.3	Unspecified	Crutzen *et al.* (1979)
	5–15	Unspecified	Stocks (2004)
	8.5	1940–1991	Kasischke *et al.* (1993)
	12	Average	Levine and Cofer (2000)
North America	3	1980–1999	Stocks (2004)
	<5 (with extremes of ≈7)	Average	Goldammer and Furyaev (1996b)
Russia	2.3 (with extremes of 16)	Average	Goldammer and Furyaev (1996b)
	2.9	1971–1991	Dixon and Krankina (1993)
	4.5	1940–1991	Kasischke *et al.* (1993)
	5–8	1990s	Shivedenko and Nilsson (2000)
	12	Average	Conard and Ivanova (1997)
Fennoscandia	<0.004 (with extremes of 0.030)	Average	Goldammer and Furyaev (1996b)
Northeast China	<0.05 (with extremes of 1.30)	Average	Goldammer and Furyaev (1996b)

et al. 2004). It will be possible to obtain accurate boreal wildfire information, even from the remotest areas and within short intervals, without having to rely solely on statistics provided by government agencies.

Ecological significance

Goldammer and Furyaev (1996a) stated that boreal wildfires are the most important natural factor controlling forest age structure, species composition, and physiognomy, as well as landscape-scale diversity, and influencing energy flows and biogeochemical cycles. More recently, boreal wildfires have gained a prominent role in discussions of ecological processes at a global scale, most notably in relation to carbon budgets and climate change (Kasischke and Stocks 2000b). This is because boreal wildfires consume 25–50 t of fuel per hectare, which is more than four times the fuel consumption by wildfires in the savanna

INTRODUCTION

biome, even though the boreal biome has a smaller annual extent of wildfires (Levine and Cofer 2000). As a result, boreal wildfires release more than 180 Mt of carbon annually (van der Werf *et al.* 2010).

Many regional-scale ecological processes are also related to boreal wildfires. These include mass migrations of pyrophilic animal species that are attracted to post-fire sites (Nappi and Drapeau 2009), perpetuation of forest succession patterns that require periodic wildfires to reset the species-composition patterns (Payette 1992), and predator–prey relationships that are linked to periodic fire disturbance (Fox 1983). Boreal wildfires also influence local ecological processes by, for example, assisting the regeneration of certain tree species and releasing nutrients that are held in the slow-decomposing litter layer (MacLean *et al.* 1983). In fact, in some parts of the boreal forest biome, where wildfires have been eliminated by land conversion and fire suppression (Parviainen 1996), it is considered necessary to encourage wildfires because of their importance for ecological processes (Pyne 1996).

Suppression of fire is a major focus of boreal resource management agencies. Major motivations for the elimination of fires and reductions in fire size have been the need to protect people, avoid damage to infrastructure, and minimize the loss of standing timber (Ward and Mawdsley 2000). Even though eradication of damage from wildfires to protect societal and economic values is a cultural and political necessity, fire suppression policies and actions are not without ecological consequences. Commonly mentioned ecological effects of aggressive fire suppression include excessive fuel buildup that leads to extreme fire hazards, and alterations of ecosystem patterns and processes that lead to a loss of biodiversity (Grissom *et al.* 2000).

Major components of the ecological patterns and processes that are associated with boreal wildfires include the live and dead biomass left behind within boreal wildfires, which are broadly termed "residuals" and form the subject of this book.

Goals and scope of the book

The overall goal of this book is to summarize what is known about the ecology of post-wildfire residuals in boreal forests. By *ecology*, we mean the biological, physical, and chemical processes that are associated with the formation, distribution, and functions of the residuals, including the effects of the residuals on other organisms. By *wildfire*, we mean the physical process that creates residuals, typically large and intense fires. By *residuals*, we mean the remnant vegetation after wildfires, including dead, dying, and live plants. By *boreal forests*, we mean the northernmost circumpolar forest biome.

INTRODUCTION

Our specific goals are to:

- create awareness about boreal wildfire residuals and their ecology;
- critically examine, summarize, and synthesize the present state of knowledge; and
- explore future knowledge requirements and suggest ways to improve that knowledge.

There are caveats. We do not intend this book to be a detailed or exhaustive review of wildfires and all ecological processes that occur in post-fire forest ecosystems. Also, we mostly discuss ecological processes from a tree-centric view, which may exclude many finer-scale processes associated with lesser plants, with microorganisms, and with animal species. We also exclude from our discussion the effects of planned forest fires, such as prescribed burns, and low-intensity surface fires that do not affect or alter the tree canopy layer.

Although we address wildfire residuals and their ecology from a pan-boreal perspective, most of our discussion will be based on examples from North America and Fennoscandia. This is not by design, but is rather the result of a paucity of English-language literature on boreal forest fires and residuals from Eastern Europe, Siberia, and northern Asia, where boreal forests exist and wildfires are common. When information for boreal forests was not available, we occasionally relied on information from other similar temperate forest ecosystems. However, we believe that the breadth and the depth of topics addressed and the examples used here provide an adequate and generalized description of the ecology of residuals in pan-boreal wildfires.

This book is organized according to the flow of topics illustrated in Figure 1.2. The purpose of this first chapter has been to provide background information about boreal forests and boreal wildfires. In Chapter 2, we describe how the interaction of boreal forest vegetation and boreal wildfires form residuals. Subsequently, we describe and define various types of boreal wildfire residuals in Chapter 3 and provide information about their known extents in the boreal forest. In Chapter 4, we examine the many ecological roles of boreal wildfire residuals, for different types of residuals and different ecological processes across a range of scales. In Chapter 5, we discuss several major forest management applications that have gained importance through increasing knowledge of the ecological value of wildfire residuals.

In all of these chapters, we not only briefly review the state of knowledge and attempt to synthesize it, but also examine the major knowledge gaps and uncertainties and discuss the implications of what is not known. Whenever possible, we offer suggestions to solve these problems. In the final chapter, we summarize and synthesize the overall state of knowledge and the major knowledge gaps and uncertainties, and we identify future directions for research to improve the knowledge base about boreal wildfire residuals.

INTRODUCTION

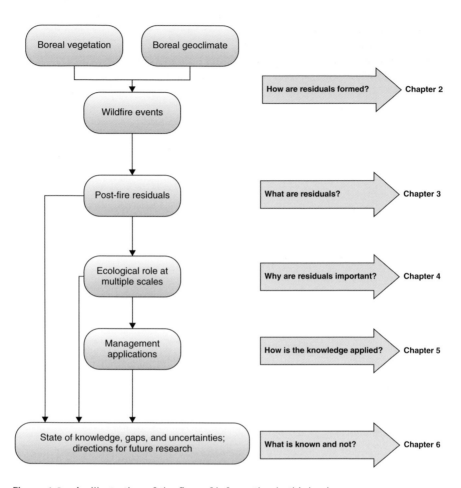

Figure 1.2   An illustration of the flow of information in this book.

INTRODUCTION

References

Ahern, F.J., Epp, H., Cahoon, D.R. Jr,, French, N.H.F., Kasischke, E.S., and Michalek, J.L. (2000) Using visible and near-infrared satellite imagery to monitor boreal forests. In Kasischke, E.S. and Stocks, B.J. (eds) Fire, Climate Change, and Carbon Cycling in the Boreal Forest. Springer, New York, Ecological Studies 138. pp. 312–330.

Alexeyev, V.A., Birdsey, R.A., Stakanov, V.D., and Korotkov, I. A. (2000) Carbon storage in the Asian boreal forest of Russia. In Kasischke, E.S. and Stocks, B.J. (eds) Fire, Climate Change, and Carbon Cycling in the Boreal Forest. Springer, New York, Ecological Studies 138. pp. 239–257.

Apps, M.J., Kurz, W.A., Luxmoore, R.J. et al. (1993) Boreal forests and tundra. Water, Air and Soil Pollution 70, 39–53.

Bonan, G.B. and Shugart, H.H. (1989) Environmental factors and ecological processes in boreal forests. Annual Review of Ecology and Systematics 20, 1–28.

Bourgeau-Chavez, L.L., Alexander, M.E., Stocks, B.J., and Kasischke, E.S. (2000) Distribution of forest ecosystems and the role of fire in the North American boreal region. In Kasischke, E.S. and Stocks, B.J. (eds) Fire, Climate Change, and Carbon Cycling in the Boreal Forest. Springer, New York, Ecological Studies 138. pp. 111–131.

Brandt, J.P. (2009) The extent of the North American boreal zone. Environmental Reviews 17, 101–161.

Bryson, R.A. (1966) Air masses, streamlines, and the boreal forest. Geographical Bulletin 8, 228–269. (Cited in Bonan and Shugart 1989.)

Conard, S.G. and Ivanova, G.A. (1997) Wildfire in Russian boreal forests – potential impacts of fire regime characteristics on emissions and global carbon balance estimates. Environmental Pollution 98, 305–313.

Crutzen, P.J., Heidt, L.E., Krasnec, J.P., Pollock, W.H., and Seiler, W. (1979) Biomass burning as a source of atmospheric gases CO, H_2, N_2O, CH_3Cl and COS. Nature 282, 253–256.

de Groot, W.J., Cantin, A.S., Flannigan, M.D., Soja, A.J., Gowman, L.M., and Newberry, A. (2013) A comparison of Canadian and Russian boreal forest fire regimes. Forest Ecology and Management 294, 23–34.

Dixon, R.K. and Krankina, O.N. (1993) Forest fires in Russia, carbon dioxide emissions to the atmosphere. Canadian Journal of Forest Research 23, 700–705.

Esseen, P.-A., Ehnström, B., Ericson, L., and Sjöberg, K. (1997) Boreal forests. Ecological Bulletins 46, 16–47.

FAO (2001). Global forest resource assessment 2000. Main Report. United Nations Food and Agriculture Organization, Rome. FAO For. Paper 140. Available from http://www.fao.org/docrep/004/y1997e/y1997e00.htm (accessed May 2013).

Fox, J.F. (1983) Post-fire succession of small mammal and bird communities. In Wein, R.W. and MacLean, D.A. (eds) The Role of Fire in Northern Circumpolar Ecosystems. John Wiley & Sons, Ltd, New York, Scope 18. pp. 155–180.

Goldammer, J.G. and Furyaev, V.V. (eds) (1996a) Fire in Ecosystems of Boreal Eurasia. Kluwer Academic Publishers, Dordrecht, Forestry Sciences Vol. 48.

Goldammer, J.G. and Furyaev, V.V. (1996b) Fire in ecosystems of boreal Eurasia: ecological impacts and links to the global system. In Goldammer, J.G. and Furyaev, V.V. (eds) Fire in Ecosystems of Boreal Eurasia. Kluwer Academic Publishers, Dordrecht, Forestry Sciences Vol. 48. pp. 1–20.

Grissom, P., Alexander, M.E., Cella, B. et al. (2000) Effects of climate change on management and policy: mitigation options in the North American boreal forest. In Kasischke, E.S. and Stocks, B.J. (eds) Fire, Climate Change, and Carbon Cycling in the Boreal Forest. Springer, New York, Ecological Studies 138. pp. 85–103.

Gromtsev, A. (2002) Natural disturbance dynamics in the boreal forests of European Russia: a review. Silva Fennica 36, 41–55.

Hytteborn, H., Maslov, A.A., Nazimova, D.J., and Rysin, L.R. (2005) Boreal forests of Eurasia. In Andersson, F. (ed.) Coniferous Forests. Elsevier, Amsterdam, Ecosystems of the World 6. pp. 23–99.

Johnson, E.A. (1992) Fire and Vegetation Dynamics: Studies from the North American Boreal Forest. Cambridge University Press, Cambridge, UK.

Kasischke, E.S. (2000) Boreal ecosystems in the global carbon cycle. In Kasischke, E.S. and Stocks, B.J. (eds) Fire, Climate Change, and Carbon Cycling in the Boreal Forest. Springer, New York, Ecological Studies 138. pp. 19–30.

Kasischke, E.S., Christensen, N.L. Jr, and Stocks, B.J. (1995) Fire, global warming, and the carbon balance of boreal forests. Ecological Applications 5, 437–451.

Kasischke, E.S., French, N.H.F., Harrell, P., Christensen, N.L. Jr, Ustin, S.L., and Barry, D. (1993) Monitoring of wildfires in boreal forests using large area AVHRR NDVI composite image data. Remote Sensing of Environment 45, 61–71.

Kasischke, E.S. and Stocks, B.J. (eds) (2000a) Fire, Climate Change, and Carbon Cycling in the Boreal Forest. Springer, New York, NY. Ecological Studies 138.

Kasischke, E.S. and Stocks, B.J. (2000b) Introduction. In Kasischke, E.S. and Stocks, B.J. (eds) Fire, Climate Change, and Carbon Cycling in the Boreal Forest. Springer, New York, Ecological Studies 138. pp. 1–6.

Krebs, J.S. and Barry, R.G. (1970) The Arctic front and the tundra-taiga boundary in Eurasia. Geographical Reviews 60, 548–554.

Kuusela, K. (1992) The boreal forests: an overview. Unasylva 43(170). Available from http://www.fao.org/docrep/u6850e/u6850e00.htm (accessed May 2013).

Larsen, J.A. (1980) The Boreal Ecosystem (Physiological Ecology). Academic Press, New York, NY.

Levine, J.S. and Cofer, W.R. (2000) Boreal forest fire emission and the chemistry of the atmosphere. In Kasischke, E.S. and Stocks, B.J. (eds) Fire, Climate Change, and Carbon Cycling in the Boreal Forest. Springer, New York, Ecological Studies 138. pp. 31–48.

MacLean, D.A., Woodley, S.J., Weber, M.G., and Wein, R.W. (1983) Fire and nutrient cycling. In Wein, R. W. and MacLean, D. A. (eds) The Role of Fire in Northern Circumpolar Ecosystems. John Wiley & Sons, Ltd, New York, Scope 18. pp. 111–134.

Nappi, A. and Drapeau, P. (2009) Reproductive success of the black-backed woodpecker (*Picoides arcticus*) in burned boreal forests: are burns source habitats? Biological Conservation 142, 1381–1391.

NRDC (2004) The boreal forest: Earth's green crown. Natural Resources Defense Council, New York. Available from http://www.nrdc.org/land/forests/boreal/intro.asp (accessed May 2013).

Parviainen, J. (1996). The impact of fire on Finnish forests in the past and today. In Goldammer, J.G. and Furyaev, V.V. (eds) Fire in Ecosystems of Boreal Eurasia. Kluwer Academic Publishers, Dordrecht, Forestry Sciences Vol. 48. pp. 55–64.

Payette, S. (1992) Fire as a controlling process in the North American boreal forest. In Shugart, H.H., Leemans, R., and Bonan, G.B. (eds) A Systems Analysis of the Global Boreal Forest. Cambridge University Press, New York. pp. 144–169.

Pyne, S.J. (1996) Wild hearth: a prolegomenon to the cultural fire history of northern Eurasia. In Goldammer, J.G. and Furyaev, V.V. (eds) Fire in Ecosystems of Boreal Eurasia. Kluwer Academic Publishers, Dordrecht, Forestry Sciences Vol. 48. pp. 21–44.

Rumney, G.R. (1968) Climatology and the World's Climates. MacMillan, New York. (Cited in Bonan and Shugart 1989.)

Scotter, G.W. (1972) Fire as an ecological factor in boreal forest ecosystems of Canada. In Fire in the Environment: Symposium Proceedings. USDA Forest Service, Intermountain Forest and Range Experiment Station, Ogden, UT. Report No. FS-276. pp. 15–25.

Shivedenko, A.Z. and Nilsson, S. (2000) Fire and the carbon budget of Russian forests. In Kasischke, E.S. and Stocks, B.J. (eds) Fire, Climate Change, and Carbon Cycling in the Boreal Forest. Springer, New York, Ecological Studies 138. pp. 289–311.

Shorohova, E., Kuuluvainen, T., Kangur, A., and Jogiste, K. (2009) Natural stand structures, disturbance regimes and successional dynamics in the Eurasian boreal forests: a review with special reference to Russian studies. Annals of Forest Science 66, 1–20.

Stocks, B.J. (2004) Forest fires in the boreal zone: climate change and carbon implications. UNECE Timber Committee/FAO European Forestry Commission. International Forest Fire News 31, 122–131.

Stocks, B.J., Mason, J.A., Todd, J.B. et al. (2003) Large forest fires in Canada, 1959–1997. Journal of Geophysical Research 108(DI), 1–12.

Sukhinin, A.I., French, N.H.F., Kasischke, E.S. et al. (2004) AVHRR-based mapping of fires in Russia: new products for fire management and carbon cycle studies. Remote Sensing of Environment 93, 546–564.

UNECE (2000) Forest Resources of Europe, CIS, North America, Australia, Japan and New Zealand. Main Report. UNECE Contribution to the Global Forest Resource Assessment. United Nations Economic Commission for Europe, Geneva.

Van Cleve, K. and Dyrness, C.T. 1983. Introduction and overview of a multidisciplinary research project: the structure and function of a black spruce, Picea mariana, forest in relation to other fire-affected taiga ecosystems. Canadian Journal of Forest Research 13, 695–702.

van der Werf, G.R., Randerson, J.T., Giglio, L. et al. (2010) Global fire emissions and the contribution of deforestation, savanna, forest, agricultural, and peat fires (1997–2009). Atmospheric Chemistry and Physics 10, 11707–11735.

Ward, P.C. and Mawdsley, W. (2000) Fire management in the boreal forests of Canada. In Kasischke, E.S. and Stocks, B.J. (eds) Fire, Climate Change, and Carbon Cycling in the Boreal Forest. Springer, New York, Ecological Studies 138. pp. 66–84.

Wein, R.W. and MacLean, D.A. (eds) (1983a) The Role of Fire in Northern Circumpolar Ecosystems. John Wiley & Sons, Ltd, New York, Scope 18.

Wein, R.W. and MacLean, D.A. (1983b) An overview of fire in northern ecosystems. In Wein, R.W. and McLean, D.A. (eds) The Role of Fire in Northern Circumpolar Ecosystems. John Wiley & Sons, Ltd, New York, Scope 18. pp. 1–18.

INTRODUCTION

2 Formation of wildfire residuals

What factors may explain the occurrence of post-fire residual vegetation in boreal forests? That is the question we explore in this chapter in the context of boreal forest communities and boreal wildfires. Even though the boreal biome includes a wide range of geographical and climatic conditions across three continents, some commonalities exist in the pan-global boreal forest communities, terrain, ecological processes, and major natural disturbances (Shugart *et al.* 1992). For example, as we described in Chapter 1, wildfires burn a large total area annually throughout this biome, and the frequency of wildfire occurrence is high. Furthermore, boreal wildfires can be intense, their individual spatial signatures can be large, and they almost always contain residual vegetation. Our focus here is the formation of those wildfire residuals. We explore the many types of residual vegetation, their extent, and their spatial distribution in the next chapter.

Any discussion of the formation of wildfire residuals in boreal forest landscapes must involve three topics: the forest communities that sustain the wildfires by providing fuel, and that eventually constitute the wildfire residuals; the wildfire behavior and the thermal energy generated during the fire, which causes mortality of some or all of the vegetation; and finally, the spatial patterns of the wildfire residuals.

To address these topics, we have organized this chapter to answer two major questions: what factors affect the formation of wildfire residuals, and how are

Ecology of Wildfire Residuals in Boreal Forests, First Edition. Ajith H. Perera and Lisa J. Buse.
© 2014 Ajith H. Perera and Lisa J. Buse. Published 2014 by John Wiley & Sons, Ltd.
Companion Website: www.wiley.com/go/perera/wildfire

those residuals formed and spatially distributed during wildfires? Our treatment of boreal forest communities and wildfire processes will provide only an introduction to these topics, and is therefore not exhaustive or authoritative. Indeed, such an exercise is beyond the scope of this book. To learn more, we recommend several excellent texts that provide a more comprehensive discussion of the topic of boreal vegetation (e.g., Larsen 1980, Scott 1995) and boreal wildfires (e.g., Wein and MacLean 1983, Johnson 1992, Goldammer and Furyaev 1996). Also, we caution that much variability exists across the boreal regions in the specific details of both vegetation and fire events. Our attempt here may therefore appear to be an oversimplification and generalization. We hope that readers will understand that, rather than trying to account for all possible exceptions, we have instead tried to find generalities that are valid across the boreal region.

Factors that affect the formation of residuals

To understand how wildfire residuals form, familiarity with the types of boreal forest community and the behavior of wildfires is required. The former provides insights into the characteristics of the combustible material that will sustain a wildfire and the characteristics of the potential residuals. The latter provides insights into how the residuals form and how they are spatially distributed within a *wildfire footprint* – the spatial boundary that encompasses all fire processes. By addressing only boreal forest communities and wildfire behavior, we do not neglect the importance of climate, weather, and the geo-environment (e.g., geology,

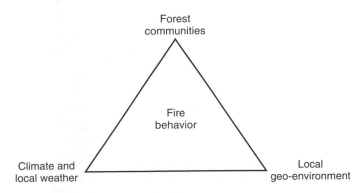

Figure 2.1 An abstract depiction of the major factors that affect the formation of residuals in boreal wildfires. These factors operate at different ecological scales. For example, long-term climate and the local geo-environment produce changes in the types of forest communities, whereas short-term local weather and the local geo-environmental conditions directly affect the behavior of a given fire.

FORMATION OF WILDFIRE RESIDUALS

topography, soil moisture regime) in forming residuals. For example, they may change fire behavior or modify the factors that determine the composition of the residuals (Figure 2.1). Indeed, we address climate and geo-environmental variables when we discuss residual formation and the factors that influence this process.

Boreal forest communities

Boreal forests are characterized by relatively few tree species that dominate the vegetation. These species belong to eight genera that are common throughout the range of boreal forests around the world: the conifers spruce (*Picea*), fir (*Abies*), pine (*Pinus*), and larch (*Larix*) and the hardwoods poplar (*Populus*), birch (*Betula*),

Table 2.1 Major tree species in the world's boreal forests and their longitudinal limits (adapted from Larsen 1980).

Genus	Continent (range of longitudes)			
	North America (55–160°W)	Northern Europe (5–40°E)	Western Siberia (40–120°E)	Eastern Siberia (120–170°W)
Conifers				
Picea (spruce)	*glauca* *mariana*	*abies* (syn. *excelsa*)	*obovata*	*obovata* *jezoensis*
Abies (fir)	*balsamea* *lasiocarpa*		*sibirica*	*nephrolepis* *sachalinensis*
Pinus (pine)	*banksiana* *contorta*	*sylvestris*	*sylvestris* *sibirica*	*sylvestris* *pumila* *cembra*
Larix (larch)	*laricina*		*sibirica* *sukaczewii*	*gmelinii* (syn. *dahurica*)
Hardwoods				
Populus (poplar)	*tremuloides* *balsamifera*	*tremula*	*tremula*	*tremula* *suaveolens*
Betula (birch)	*papyrifera* *kenaica*	*pubescens* *pendula* (syn. *verrucosa*) *kusmisscheffii*	*pubescens* *pendula* (syn. *verrucosa*)	*ermanii* *platyphylla*
Alnus (alder)	*incana* (ssp. *tenuifolia*) *viridis* (ssp. *crispa*, ssp. *rugosa*)	*incana* *viridis* (ssp. *crispa*)	*viridis* (ssp. *fruticosa*)	*viridis* (ssp. *fruticosa*)
Salix (willow)	*Salix* spp.	*Salix* spp.	*Salix* spp.	*Salix* spp.

FORMATION OF WILDFIRE RESIDUALS

alder (*Alnus*), and willow (*Salix*). The dominant species in each genus change from region to region, but a given species typically has a broad geographical range within its continent. Larsen (1980) summarized their distribution based on longitude (Table 2.1).

Within these broad geographic distribution patterns, the regional distribution of tree species and their associated vegetation communities change in response to variations in geology, local climate, and topography, as has been documented for North America (e.g., Larsen 1980, Scott 1995), Fennoscandia (e.g., Esseen *et al.* 1997), and Siberia (e.g., Wein and MacLean 1983). One main factor that determines the local occurrence of tree species appears to be soil moisture (Table 2.2), and the direct effects of other factors, such as soil nutrient content, may be less evident (Larsen 1980). Both Wein and MacLean (1983) and Esseen *et al.* (1997) noted that in boreal forests in general, pines dominate the drier sites, and spruces and firs dominate the moister sites.

Hardwood species are also associated with certain soil moisture regimes, though the association appears weaker than it is for the conifers. For example, in North America, white birch (*Betula papyrifera*), balsam poplar (*Populus balsamifera*), and trembling aspen (*Populus tremuloides*) occur on sites ranging from very dry to wet, although balsam poplar grows mostly on moister sites (Larsen 1980).

Many boreal forest communities are centered on these tree species; that is, distinct assemblages develop in which the major tree species are associated with certain shrub, herb, bryophyte, and lichen species. Details of their composition, how they form, and where they occur have been studied for a long time in North

Table 2.2 Association of the major coniferous species with soil moisture regimes across the major boreal forest regions (summarized from Larsen 1980, Wein and MacLean 1983, and Esseen *et al.* 1997).

Soil moisture regime	Region				
	Western North America	Eastern North America	Northern Europe	Western Siberia	Eastern Siberia
Dry	*Pinus contorta*	*Pinus banksiana*	*Pinus sylvestris*		*Pinus cembra*
Moist	*Picea mariana* *Picea glauca*	*Picea mariana* *Picea glauca* *Abies balsamea*	*Picea abies*	*Picea abies*	*Picea obovata*
Wet	*Picea mariana*	*Larix laricina* *Picea mariana*	*Picea abies* *Larix* spp.	*Picea obovata* *Larix sibirica* *Abies sibirica* *Larix sukaczewii* *Pinus sibirica*	*Larix gmelinii* *Pinus sibirica*

FORMATION OF WILDFIRE RESIDUALS

America, Fennoscandia, and Eurasia. Such findings are mentioned regularly in the scientific literature and are readily available to interested readers. However, most of these descriptions address local variations rather than being attempts to generalize spatial patterns, and are therefore not useful in articulating the broad patterns that we focus on in this book. For readers who are interested in more detailed information about boreal forest communities, we highly recommend Larsen (1980) and Scott (1995).

In the rest of this section, we present a simplified and generalized interpretation of the occurrence of boreal forest communities based on a combination of the parent geological material, soil type, topography, and water table characteristics. This overview is based on North American examples, but Scott (1995) suggests that parallel boreal forest communities exist in northern Europe, Siberia, and the Far East, including northern China and Japan. Figure 2.2, adapted from Scott (1995), illustrates four major types of boreal forest communities, their associations with geo-environmental conditions, and their juxtaposition. We reiterate that, in addition to local exceptions, these associations may be altered and further complicated by repeated disturbance, such as by fire, insect defoliation, or forest harvesting. Our focus is on the generalities that influence the susceptibility of boreal forest communities to burning and that define the characteristics of the wildfire residuals.

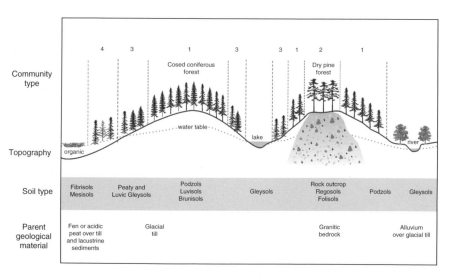

Figure 2.2   A broad view of the occurrence and juxtaposition of mature seral stages of boreal forest communities in relation to geo-environmental conditions (adapted from Scott 1995). The major community types are: (1) closed coniferous forest; (2) dry pine forest; (3) lowland spruce; and (4) spruce–larch wetland. In types (1) and (3), upland hardwood and mixedwood stands can occur as transitional phases during succession.

We have provided general descriptions of these main broad boreal forest community types, based on the more detailed descriptions given by Larsen (1980) and Scott (1995) for North America, Scott (1995) for Eurasia, and Esseen *et al.* (1997) for Fennoscandia. These descriptions focus on the mature seral stage of conifer-dominated communities. Missing from these broad conifer types is an open spruce–lichen woodland condition, which occurs at the northern edge of the boreal forest where it transitions into tundra. We mention this northernmost type here because it is relevant in the context of fire, as will be discussed later in this chapter (e.g., Table 2.3). In addition to the simplified coniferous forest conditions, we describe hardwood and mixedwood forest conditions, which are primarily transitional or early successional community types in the boreal biome. At any given time, all seral stages of development occur in a landscape, adding to the heterogeneity of the forest composition and structure (Figure 2.3).

Closed coniferous forest

These upland forests are typically relatively dense, closed-canopy stands of conifers in pure or mixedwood stands (Figure 2.4). In North America, the main overstory species in this stand type are white spruce (*Picea glauca*), black spruce (*Picea mariana*), and balsam fir (*Abies balsamea*). These can occur as large tracts of a single species or of combinations of species, and may be interspersed with white birch and trembling aspen on drier sites, or balsam poplar on moist sites. In drier areas, pockets of jack pine (*Pinus banksiana*) may also occur. Typically, closed coniferous forests are found on well-drained soils such as Podzols on the Canadian Shield, and on Luvisols and Brunisols. In mature forests, trees can range from 15 to 30 m tall, but average from 20 to 25 m, and typically range from 20 to 50 cm in diameter (Weber and van Cleve 2005). These closed-canopy forests are often cool and dark, allowing little light to penetrate to the forest floor, but are rarely dense enough to completely eliminate the understory (Scott 1995).

In these upland coniferous forests, the understory species vary depending on local site conditions and differ between pure and mixed stands, with greater shrub diversity occurring in mixed-species stands. Common understory shrub species in North America include Labrador tea (*Ledum groenlandicum*), *Viburnum* species, *Ribes* species, and alder (*Alnus* spp.) (Larsen 1980). The forest floor is often a combination of feathermoss cover (*Pleurozium schreberi, Hylocomium splendens, Ptilium crista-castrensis,* and *Dicranum* species) (Johnson 1992, Scott 1995), ground-dwelling *Cladonia* lichens, and, on moister sites, some sphagnum (*Sphagnum* spp.) mosses (Figure 2.5).

In this community type in Fennoscandia, the dominant tree species is Norway spruce (*Picea abies*), with a shrub-poor understory that may contain some willow (*Salix* spp.) and juniper (*Juniperus* spp.) species but that consists mostly

FORMATION OF WILDFIRE RESIDUALS

FORMATION OF WILDFIRE RESIDUALS

Table 2.3   Canadian FBP system fuel types common to boreal forest landscapes (adapted from Alexander et al. 1984 and Hirsch 1996) and their associated forest community types.

Fuel type	Stand structure and composition	Forest floor	Organic layer	Surface fuels			Ladder fuels	General forest community type associations
				Herbs and shrubs	Understory conifers	Dead and downed fuels		
Spruce–lichen woodland C-1	Open stands of black spruce in dense clumps on well-drained upland sites. Jack pine and white birch can occur in small amounts	Continuous reindeer lichen (mainly *Cladonia*) 3 to 4 cm deep over mineral soil	Shallow (0–5 cm) or absent; uncompacted	Very sparse	Absent	Light and scattered	Live and dead black spruce branches extending to the forest floor; layering is extensive	Closed coniferous (in geographic transition)
Boreal spruce C-2	Moderately well-stocked black spruce stands on upland and lowland sites (excluding sphagnum bogs)	Continuous feathermoss and ground-dwelling reindeer lichen (usually *Cladonia*)	Compact and deep (often exceeding 20–30 cm)	Continuous Labrador tea is often dominant	Scattered in open stands; absent in denser stands	Low to moderate	Tree crowns extend to or near the ground; tree bark is flaky, and arboreal lichens may be present on the branches	Closed coniferous and lowland spruce
Mature jack or lodge-pole pine C-3	Fully stocked (1000–2000 stems ha^{-1}) mature jack pine or lodgepole pine stands	Continuous feathermoss	Compact and moderately deep (about 10 cm)	Sparse to moderate	Absent or sparse	Light and scattered	Absent; the base of the live crown is well above the surface fuels	Dry pine
Immature jack or lodge-pole pine C-4	Pure, dense stands (10 000–30 000 stems ha^{-1}) of immature jack pine or lodgepole pine	Continuous needle litter	Shallow (0–5 cm) and moderately compact	Moderate cover; a low shrub layer (e.g., *Vaccinium* spp.) is often present	Large quantity of standing dead stems	Heavy	Continuous vertical and horizontal continuity	Dry pine

FORMATION OF WILDFIRE RESIDUALS

Leafless aspen D-1	Pure, semi-mature, moderately well-stocked stands of trembling aspen	Continuous leaf litter	Shallow (0–5 cm) and uncompacted	Well-developed medium to tall shrub layer; moderate herbaceous layer	Absent	Sparse	Absent	Upland hardwood and mixedwood
Boreal mixedwood M-1: leafless M-2: green	Moderately well-stocked stands of boreal coniferous species (e.g., black or white spruce, balsam fir) and deciduous species (e.g., aspen, white birch). Variations occur with season and with the proportions of coniferous and deciduous trees	Continuous leaf litter in deciduous portions of the stands. Discontinuous feathermoss and needle litter in coniferous portions	Shallow (0–5 cm); uncompacted to moderately compacted	Moderate shrub and continuous herb layer	Scattered to moderate	Low to moderate	Conifer crowns may extend to near the ground in some stands	Closed coniferous and upland hardwood and mixedwood
Dead balsam fir mixedwood M-3: leafless M-4: green	Moderately well-stocked stands of black or white spruce, jack pine, or white birch with dead balsam fir (often as an understory). Variations in type occur between seasons	Continuous leaf litter in deciduous portions of stands. Discontinuous feathermoss, needle litter, and hardwood leaves in mixed portions	Moderately compact and moderately deep (8–10 cm)	Dense continuous herbaceous cover after green-up occurs	Dead balsam fir	Initially low, becoming heavy 5–8 years after balsam fir mortality occurs	Dead balsam fir understory with branches covered by arboreal lichen	Closed coniferous and upland hardwood and mixedwood

Figure 2.3   An example of the boreal forest landscape: relatively flat terrain interspersed with lakes, wetlands, and forest communities of different composition and ages. The spatial mosaic of forest communities is constantly in flux because of frequent wildfires.

Figure 2.4   An upland spruce forest; this forest type is a major component of the boreal landscape.

Figure 2.5   An upland spruce stand with a moderate shrub understory and a mixture of forest-floor species, illustrating closed canopy conditions.

of blueberries and lingonberries (*Vaccinium* spp.). The forest floor is dominated by feathermosses such as *H. splendens* and *P. schreberi* (Esseen *et al.* 1997). In Europe, this community type is dominated by spruce and in eastern Siberia by larch species (Scott 1995), again with an understory that has few shrubs or herbs and a forest floor that is covered in feathermosses (Larsen 1980). Mature trees can range from 30 to 45 m tall and up to 1 m in diameter (Hytteborn *et al.* 2005).

Toward the northern edge of the North American boreal forest, tree heights and diameters gradually decrease, so that these coniferous stands are more open, and become what is often referred to as open spruce–lichen woodland (Scott 1995, Weber and van Cleve 2005). In Eurasia, these open woodlands are dominated by larch (*Larix gmelinii*), especially in western Siberia, with a minimal understory of dwarf shrubs and lichens (Scott 1995).

Dry pine forest

The dry pine community type is very common in boreal forests, where it is found on Regosols, sandy outwash plains, old beaches, and granitic outcrops. This xeric forest community is dominated by jack pine in eastern North America (Figure 2.6) and by lodgepole pine (*Pinus contorta*) in the west. Mature stands are between 17 m and 20 m tall (Weber and van Cleve 2005) and can range from

FORMATION OF WILDFIRE RESIDUALS

(a)

(b)

Figure 2.6   Examples of jack pine forest with (a) a low shrub understory and (b) a continuous feathermoss understory illustrating the variability in the density of this forest type. (Example (b) reproduced with permission from P. Uhlig, Ontario Ministry of Natural Resources.)

dense closed-canopy forests to fairly open stands, both with a minimal shrub understory. The ground flora consists of juniper, blueberries, feathermosses, and lichens (*Cladina* spp.) (Larsen 1980). At the southern edge of the boreal forest, red pine (*Pinus resinosa*) and white pine (*Pinus strobus*) can form part of the tree canopy in this community type, especially on rocky outcrops (Larsen 1980).

In Fennoscandia, Scots pine (*Pinus sylvestris*) is the dominant tree species on drier sites (Scott 1995), with an understory of dwarf shrubs such as heather (*Calluna vulgaris*) and crowberry (*Empetrum hermaphroditum*), as well as blueberries and bracken fern (*Pteridium aquilinum*), and the forest floor is covered by lichens of the genus *Cladonia* and less often *Stereocaulon* (Scott 1995, Esseen *et al.* 1997). On fertile sites, Scots pine can grow up to 30 m tall and up to 1 m in diameter (Hytteborn *et al.* 2005) but will be smaller on dry, nutrient-poor sites. In Eurasia, dry pine stands are generally open, with relatively little understory, creating almost park-like conditions.

Lowland spruce forest

The lowland spruce community type is dominated by black spruce and white birch in North America, and occurs on poorly drained Gleysols where the water table is relatively high. Balsam poplar and trembling aspen are also common in these communities, with the latter species becoming more common in western North America. Trees in these lowlands tend to be shorter (typically less than 15 m tall) and more variable in height and diameter than those growing on upland sites (Figure 2.7). This may be because the organic matter that dominates the soils of the former sites provides limited seedbed, and the spruce progressively becomes established over many decades through layering (vegetative reproduction from live branches that become buried under litter), which results in an uneven age structure (Weber and van Cleve 2005). The understory is sparse, often consisting of Labrador tea. The forest floor is typically covered by a continuous mat of sphagnum mosses (Larsen 1980). European and Fennoscandian equivalents for this forest type were not described in the sources we consulted.

Spruce–larch wetland

The spruce–larch community type in North America is dominated by black spruce and eastern larch (*Larix laricina*), with the understory consisting primarily of ericaceous shrubs such as leatherleaf (*Chamaedaphne calyculata*) and *Ledum* species, with *Carex* species in wetter areas and with a thick sphagnum mat overlying organic soils (Larsen 1980, Scott 1995). Tree cover is typically sparse and growth is restricted (Figure 2.8). In some areas, larch dominates these poorly drained organic sites and is often mixed with spruce and alder species (Larsen

FORMATION OF WILDFIRE RESIDUALS

Figure 2.7   A lowland spruce community, showing the height variability that is typical of this community type.

Figure 2.8   A spruce–larch wetland community with an understory of ericaceous shrubs, in which the open and sparse nature of this forest type is evident.

1980). In the Siberian boreal forest, the dominant species on these sites are Siberian spruce (*Picea obovata*), Siberian larch (*Larix sibirica*), and Russian larch (*Larix sukaczewii*) (Wein and MacLean 1983), with an understory of dwarf shrubs such as wild rosemary (*Ledum palustre*) and blueberries, and a forest floor covered by sphagnum mosses (Scott 1995).

Upland hardwood and mixedwood forest

In both North America and Fennoscandia, relatively species-rich mixed hardwood forests form an early to intermediate post-fire successional stage and may also represent an environmental transition from boreal forest to sub-boreal hardwood and mixedwood forests. In this community type in North America, trembling aspen can form pure stands over a range of edaphic conditions, in which the common understory shrubs comprise willow species, speckled alder (*Alnus incana* ssp. *rugosa*), red raspberry (*Rubus idaeus*), and red-osier dogwood (*Cornus stolonifera*) (Figure 2.9a). The forest floor can include a diversity of herbaceous species. Also, intermittent patches of mixed hardwood stands occur throughout the North American boreal forest that include white birch, balsam poplar, balsam fir, green alder (*Alnus viridis* ssp. *fruticosa*), and American mountain-ash (*Sorbus americana*) (Figure 2.9b) (Scott 1995). In Fennoscandia, these hardwood and mixedwood forests are dominated by birch species and Eurasian aspen (*Populus tremula*), with a diverse, species-rich understory (Scott 1995, Esseen *et al.* 1997).

Boreal wildfires

Wildfires in boreal forest landscapes, as in other landscapes, result from the combustion of biotic materials that have been rendered readily flammable by their chemical content (e.g., waxes and terpenes in conifer needles), their moisture content (e.g., the low moisture content of the litter and coarse wood due to drying), and their physical structure (e.g., the loose arrangement of the litter layer). A fire requires a source of ignition to start the process, and suitable weather conditions to establish flaming combustion; a continuous supply of fuel is then required to sustain the spread of the flames, and either the depletion of fuel or high moisture levels in the fuel eventually extinguishes the fire (Whelan 1995). Our discussion of wildfire processes in this section – of forest fuels, fire weather, and fire behavior – is mostly based on the North American literature, and particularly that from Canada, due both to our geographical familiarity and the ready availability of published materials. This should not create an undue bias, since there are more similarities in how wildfires behave in boreal forests around the world than there are differences.

(a)

(b)

Figure 2.9   Examples of (a) aspen and (b) birch mixed hardwood forests, illustrating the diversity of species composition and understory structure. (Example (b) reproduced with permission from G. Racey, Ontario Ministry of Natural Resources.)

The knowledge of boreal wildfire processes is vast and comprehensive, so that providing a complete summary of what is known on this topic would be an enormous task, well beyond the scope of this book. We caution the reader that, in our attempt to highlight the generalities, we may have oversimplified some concepts. We refer readers interested in further detail to comprehensive compilations about wildfire processes in general (e.g., Chandler *et al.* 1983, Whelan 1995, Johnson and Miyanishi 2001) and about boreal wildfires in particular (e.g., Wein and MacLean 1983, Johnson 1992, Goldammer and Furyaev 1996). Below we summarize briefly what is known about fire processes for the purpose of providing context for discussion of how these processes influence the formation of wildfire residuals.

Forest fuel types

Four general categories of fuel can be distinguished on the basis of their location within the forest's structure (van Wagner 1983): surface fuels composed of leaf litter and fine woody debris, including live but dry lichens and mosses; subsurface fuels composed of decaying organic material of varying depths and moisture contents; coarse surface fuels composed of fallen trees; and canopy fuels that consist of the foliage and branches of standing trees. These "fuel types" vary in their mass, bulk density, moisture content, and flammable chemicals. "Ladder" fuels represent fuels in these categories that connect surface materials to the trees, thereby allowing fire to climb into the canopy. All of these fuel types collectively play a significant role in determining the behavior – that is, ignition, spread, and intensity – of a wildfire. Several formal definitions are available for the term "fuel type" (e.g., CIIFFC 2003, NWCG 2012, Stacey 2012); here we use the definition "an identifiable association of fuel elements of distinctive species, form, size, arrangement, and continuity that will exhibit characteristic fire behavior under defined burning conditions" (CIIFFC 2003).

To link the forest communities described in the previous section with fire behavior in terms of the abundance and distribution of fuels, boreal scientists and forest fire managers have developed a generalized classification system for the various fuel types. The premise of this system is that the structural and compositional differences among forest communities will produce different physical and chemical characteristics in the fuel, which will in turn lead to different fire behavior. One such system that is popular in many countries is the Canadian Forest Fire Behavior Prediction System, which is known as the "FBP system" (Alexander *et al.* 1984). According to the FBP system, boreal conifer communities can be categorized into four fuel types: upland spruce, fuel type C-1; lowland spruce (excluding interspersed sphagnum moss bogs), C-2; and dry pine, classed as C-3 if the trees are mature (Figure 2.6) and C-4 if they are immature. The mixed hardwood forest community is categorized into five fuel types based on the species composition. Pure stands of poplar (*Populus* spp.) are classified as fuel type D-1, and the four additional types are based on the species mixture and their phenology. Mixedwood communities, which include both coniferous and non-coniferous species, produce fuel types M-1 when the hardwoods are leafless, or M-2 when hardwoods have leaves. If the understory consists of dead balsam fir (*Abies balsamea*), these communities produce fuel types M-3 (when the hardwoods are leafless) or M-4 (when hardwoods have leaves). Table 2.3 provides details of the overstory, understory, and forest floor characteristics in these fuel types and their corresponding forest community types (discussed in the previous section). The FBP system's fuel classification is a practical tool used by forest fire managers in boreal North America, as it generalizes the expected fire behavior in many boreal forest communities and, in conjunction with weather and topography data, allows for quantitative estimates of wildfire characteristics (Taylor *et al.* 1997).

Fire weather

The presence of fuel alone is not sufficient to cause and sustain a wildfire. Local weather plays an equal if not more important role, especially during large fires (Johnson 1992, Meyn *et al.* 2007). Continuous periods without rain, with high temperatures, high winds, low relative humidity, or a combination of these factors, lower the moisture content of the fuel and make that fuel more vulnerable to ignition. The lightning generated by intense storms provides a source of ignition. Furthermore, high winds, low relative humidity, and low rainfall sustain burning fires. These weather parameters, which directly influence wildfire processes, are collectively known as "fire weather", and the FBP system contains a fire weather system composed of multiple indices that quantify the influence of these parameters on fire behavior. Several formal definitions are available for the term "fire weather" (e.g., CIIFFC 2003, NWCG 2012); here, we use the definition "weather conditions which influence fire ignition and behavior" (NWCG 2012).

Based on the ambient temperature, relative humidity, rainfall, and wind speed, this fire weather index system can be used to derive three codes and three indices (Figure 2.10; van Wagner and Pickett 1985). The *Fine Fuel Moisture Code* (FFMC) is a numeric rating of the moisture content of the litter, which indicates the ignitability and the flammability of fine fuel on the forest floor. It responds to short-term changes in the weather, within a single day. The *Duff Moisture Code* (DMC) codifies the average moisture content of loose organic matter in the top layers of the soil, and responds more slowly to weather patterns; for example, it responds over a period of a week or longer. This code helps to estimate the potential consumption of subsurface fuel when a fire begins.

The *Drought Code* (DC) is a numeric rating of the average moisture content of the deeper, more compact organic layers and of the downed wood and logs on the forest floor, and is a useful indicator of seasonal drought effects on forest fuels

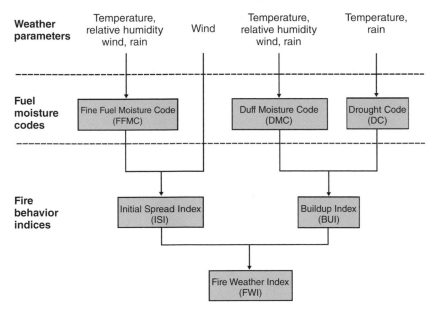

Figure 2.10   An overview of the Canadian Fire Weather Behavior Index System. The system is used to calculate three moisture codes and three fire behavior indices based on past and prevailing weather parameters for a given location. For more details, see van Wagner (1987) and http://cwfis.cfs.nrcan.gc.ca/en_CA/background/summary/fbp. (Reproduced with permission of the Minister of Natural Resources Canada, 2013.)

and the potential degree of smoldering combustion. This rating responds even more slowly to weather trends, over periods of months. The *Initial Spread Index* (ISI) is related to the expected rate of fire spread based on the FFMC and wind speed. The *Buildup Index* (BUI) is a numeric rating of the total amount of fuel available for combustion, as a function of DMC and DC. The *Fire Weather Index* (FWI) is a composite numerical rating that is widely used to indicate fire danger under given weather conditions. In combination with wind speed, topography, and fuel types, the fire weather indices of the FBP system produce quantitative estimates of the behavior and characteristics of a given wildfire (Alexander *et al.* 1984). They are especially useful for predicting the spread patterns at the onset of a fire and the fire's subsequent general behavior.

Wildfire behavior

"Fire behavior" is a broad term that has been formally defined by many authors (e.g., CIIFFC 2003, NWCG 2012, Stacey 2012); here, we adopt the CIIFFC (2003) definition, which states that fire behavior is "the manner in which fuel ignites, flame develops, and fire spreads and exhibits other related phenomena as determined by

FORMATION OF WILDFIRE RESIDUALS

the interaction of fuel, weather, and topography". An ignited fire spreads by first pre-heating the fuel that surrounds the point of ignition, through radiation, conduction, and convection, and then continues by combustion of the pre-heated fuel. This recruitment of new fuel and the ensuing fire spread are slow under conditions with little or no wind, but become rapid under high winds because of enhanced heat transfer through radiation, conduction, and convection (Figure 2.11a; Rothermel 1972). A fire that normally spreads radially outwards under a slow wind speed changes its shape from circular to elliptical when the wind speed increases, and the pattern of fire spread becomes elongated (Figure 2.11b; Fendell and Wolff 2001). As the wind speed increases further, the fire's shape becomes even more elongated; that is, it has an increasingly high length to breadth ratio (Johnson 1992).

Under homogeneous fuel and topographic conditions, as well as constant wind direction and speed, each fire contains three distinct sub-types: a head fire at the front of the burning area that spreads rapidly with the wind, flank fires on each side with lower rates of lateral spread perpendicular to the wind direction, and a backfire that spreads slowly back into the wind (Figure 2.11c; van Wagner 1969). In addition to spreading along the ground by means of the continuous combustion of surface fuels, fires can also spread through the air by a mechanism known as "spotting" (van Wagner 1983). Spotting occurs when flaming pieces of fuel become airborne during high winds and high rates of spread, then are transported to regions in front of the head fire (Figure 2.11a), where they provide multiple ignition points ahead of the main fire's path. Flaming fuel can travel several kilometers ahead of the fire front, and distances greater than 20 km have been reported (Whelan 1995).

The fire's *frontal intensity*, which represents the total heat energy transferred at the flaming fire front (CIIFFC 2003), is also referred to as the *fireline intensity* in the USA (NWGC 2012) and the *fire intensity* in Europe (Stacey 2012). As a fire spreads, this intensity is determined by the rate of spread and the mass of fuel consumed, or sometimes by the length of the flames (Johnson 1992). Typically expressed in $kW\,m^{-1}$, the frontal fire intensity in boreal wildfires can range from 10 to more than $100\,000\,kW\,m^{-1}$, with the higher values not uncommon (van Wagner 1983). van Wagner defined five categories of boreal fire based on their intensities, and these categories provide good generalized indications of fire behavior (Table 2.4). The FBP system predicts these behavioral characteristics as a function of the fuel types, fire weather indices, topography, and wind speed. It also predicts the rate of spread and fire intensity at the head, flanks, and back of the fire; fuel consumption; the classification of the fire; and the resulting area and shape (the length to breadth ratio of the ellipse) (Hirsch 1996).

It is generally agreed that most of the burned area within boreal wildfires results from the fast-spreading fire front, which transfers high amounts of heat, and from "crowning", which occurs when the flames engulf the tree canopy (the "crown"),

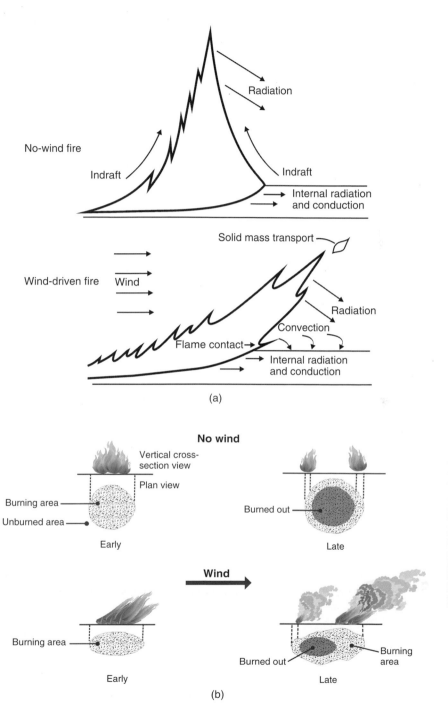

Figure 2.11   (*continued*)

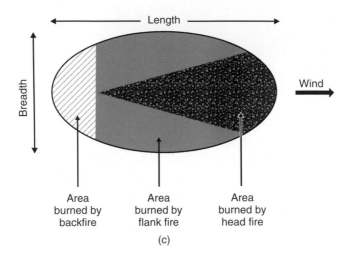

Figure 2.11   Wildfire behavior is highly sensitive to wind speed and direction. (a) By changing the flame characteristics, a wind causes the fire growth to become directional (Rothermel 1972, reproduced with permission of USFS), resulting in (b) elliptical burn areas under homogeneous fuel conditions (Fendell and Wolff 2001, reproduced with permission of Elsevier). Conceptually, such ellipses contain (c) three fire regions with different intensities and rates of spread (van Wagner 1969, reproduced with permission of Canadian Institute of Forestry).

Table 2.4   Boreal forest fire categories based on fire intensity. (van Wagner 1983. Reproduced with permission of John Wiley & Sons.)

Fire intensity (kW m^{-1})	Fire category
<10	Smoldering fire in deep organic layers
100–800	Surface backfires, burning against the wind
200–15 000	Surface head fires, burning with the wind
800–40 000	Crown fires, advancing as a single front
Up to 150 000	High-intensity spotting fires

consume the live foliage and branches, and spread rapidly from tree to tree (van Wagner 1983). There are several possible causes of crowning (Johnson 1992). One is the common occurrence of "ladder fuels". As mentioned above, these are dry and readily flammable plant materials that provide vertical continuity between fuels on the forest floor and fuels in the tree canopies, and thus provide a direct passage into the crowns for flame from surface fires. Examples of ladder fuels include trees with a conical architecture that results from retaining both live and dead branches low on the stem, dry arboreal lichens that hang from the canopy, and understory vegetation that reaches or comes close to the canopy. Another cause of crowning lies in the physical and chemical composition of boreal conifer canopies: they typically have high bulk densities (i.e., a high mass of fuel per unit of canopy volume), as well as a low moisture content, together with high levels of flammable waxes, resins, and oils in the foliage. As a rule, boreal tree species are not very tall, therefore their lower branches can be easily scorched and ignited by high-intensity surface fires (van Wagner 1983).

Furthermore, because boreal conifers are densely stocked and occur as large monocultures, fires can travel from one tree crown to another above a rapidly moving surface fire (Figure 2.12). This form of spread is very common in boreal wildfires and is termed an *active* crown fire (van Wagner 1977). In contrast, a *passive* crown fire creeps up into the canopy from the surface along ladder fuels or by direct scorching. According to Johnson (1992), active crown fires are common in boreal upland forest communities under closed-canopy conditions and passive

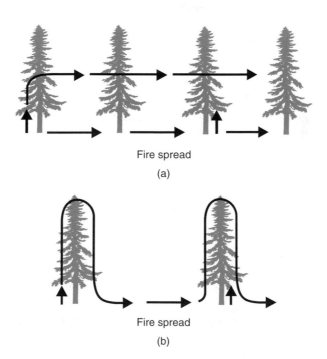

Fire spread

(a)

Fire spread

(b)

Figure 2.12   Two major types of crown fires are common in the boreal forest (based on van Wagner 1977 and Johnson 1992): (a) In an active crown fire, a high-intensity, fast-moving fire front spreads along both the surface and through the canopy simultaneously. (b) In a passive crown fire, the fire spreads along the surface, but occasionally climbs into and ignites individual trees due to low canopy heights, ladder fuels, or both. In both types of fires, tree canopies are consumed.

FORMATION OF WILDFIRE RESIDUALS

crown fires are typical in open-canopy lichen woodlands. On rare occasions, high wind speeds may allow a fire to travel momentarily and for short distances among only the tree canopies, without any surface flames.

Residual formation and distribution

The description of wildfire behavior that we have provided is an adequate generalization; that is, large boreal wildfires are ubiquitous and frequently include high-intensity crown fires that destroy most living organisms in their path (Figure 2.13).

However, much variability exists in this process because of the spatial and temporal variability of the primary factors that affect fire ignition, spread, and extinction: that is, the characteristics of the forest communities that the fire passes through, and the weather. Not only do these vary, but they do so at a multitude of spatial and temporal scales (except for topographic and geological variables that are invariant over short timeframes). In addition, the interactions between forest communities and the weather vary, resulting in a highly heterogeneous process and complex

FORMATION OF WILDFIRE RESIDUALS

Figure 2.13   A boreal wildfire often includes active crown fires that spread both along the surface and among tree canopies, with fast rates of spread and very high intensities. Even such intense fires typically produce residuals due to spatial heterogeneity in fire behavior. (Reproduced with permission of Ontario Ministry of Natural Resources.)

outcomes. Consequently, the behavior of fire within a boreal wildfire is hardly homogeneous. Significant boreal fires are characterized by spatial variability in the fire intensity, rate of spread, direction of the fire front's travel, volume of fuel consumed, and the amount of thermal energy released (e.g., Rowe and Scotter 1973, Johnson 1992). The fire behavior can change from a surface fire that has a low intensity and a low rate of spread to a major conflagration with high intensity and rapid rates of spread in a matter of minutes and meters (Jenkins *et al.* 2001). This heterogeneity in wildfire behavior within a given event is the primary reason why wildfire footprints contain residual vegetation instead of only a vast expanse of charcoal and ash. A secondary reason is the various tolerance mechanisms of plant species that enable individuals to survive the exposure to heat.

Spatial heterogeneity in fire behavior

Variability in fuel

Boreal forest landscapes are heterogeneous due to variability in topography and geology, as well as due to periodic occurrence of natural disturbances (Larsen 1980). The result is a spatial mosaic of different forest community types of differ-ent ages, and consequently a heterogeneous patchwork of fuel types (Rowe and Scotter 1973, Kafka *et al.* 2001). Even if the weather conditions remain uniform, juxtaposition of fuel types with different characteristics (Table 2.3) will cause a spreading fire to change its behavior as it moves from one type into another and will result in differences in its direction, rate of spread, flame height, and intensity. For example, under stable wind conditions, a very-high-intensity active crown fire (Table 2.4) may transform into a low-intensity surface fire when it passes over the discontinuity between fuel type C-4 and fuel type D-1. In contrast, if a spreading fire encounters C-4 on one of its flanks and D-1 at its front, the head fire direction and the associated rate of spread may change. Johnson (1992) suggested that active crown fires are more common in closed-canopy forest communities (e.g., C-2 and C-4), whereas passive crown fires are more common under open-canopy conditions (e.g., C-1). In addition, crown fires that normally require relatively high spread rates (i.e., a high ISI) can occur at low spread rates (low ISI) in fuel types such as C-4 or C-2. In addition to stand structure and composition, the age of forest stands can also affect fire behavior. For example, van Wagner (1983) hypothesized that the age of the trees can change the probability of sustaining a crown fire, with a higher probability in younger forest stands and a decreasing risk with increasing age. However, the risk of crown fire may again become higher in much older stands. This heterogeneity in fuel across the landscape creates variability in fire behavior.

FORMATION OF WILDFIRE RESIDUALS

Variability in topography

Variability in terrain can alter fire behavior even within the same fuel type. For example, steep slopes accelerate fire spread uphill by bringing the flame front closer to the surface, thereby increasing preheating of the fuel; they can thus cause higher intensities just as high wind speeds would do (Figure 2.14). Variability in the steepness of slopes and the occurrence of ridges can therefore change fire intensities even if neither the fuel type nor the wind speed changes. Conversely, fires spread more slowly downhill along steep slopes, especially as the water table approaches the surface toward valley bottoms, leaving fuels with a higher moisture content (Whelan 1995). Even in gently rolling terrain, convex surfaces such as domes that are only a few meters high are drier and produce different vegetation and fuel types (Figure 2.2); hence, they burn more rapidly than concave surfaces, which can be moister (Rowe and Scotter 1973). Open bodies of water such as lakes and areas with a high water table such as wetlands and riparian zones are ubiquitous in boreal forest landscapes. Their interspersion in expanses of forest fuel creates barriers to spreading fires because they can stop a fire front or alter its direction of spread, or, as in the case of a wetland, can reduce a fire's rate of spread (Rowe and Scotter 1973). Such topographical variations cause

FORMATION OF WILDFIRE RESIDUALS

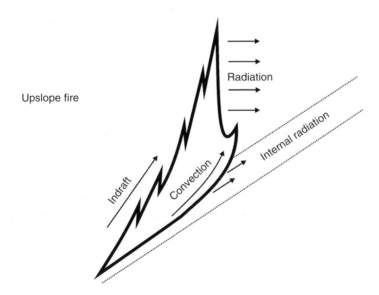

Figure 2.14   Steeper slopes assist fire spread uphill, and can compensate for a lack of wind. (Rothermel 1972. Reproduced with permission of USFS.)

a spreading wildfire to behave variably, thereby causing heterogeneous spatial patterns in frontal fire intensity.

Variability in wind and micro-weather patterns

Temporal changes in the local weather, particularly wind, can influence the variability in fire behavior within a given area. Although isolating the specific effects of wind on fire spread can be complicated (e.g., Fendell and Wolff 2001), it is generally accepted that changes in the speed and direction of wind alter the direction of the fire front's travel, its rate of spread, flame height, and fire intensity (Johnson 1992, Whelan 1995). Because wind changes both spatially and temporally during a wildfire, the fire's behavior would change accordingly, adding to the spatial heterogeneity in fire intensity caused by the variability in fuel types and topography. Local variability in rainfall prior to and during a wildfire changes fuel moisture levels, creating spatial variability in FFMC and DMC, thereby locally altering ISI and BUI (Figure 2.10). In addition, wind speeds and ambient temperature change diurnally, affecting both the rate of spread and the frontal intensity of a growing fire. It is common to observe that both the frontal fire intensity and the rate of spread peak mid-afternoon and decrease in the early morning. Sometimes these diurnal changes in frontal intensity could be by a factor of 10–20 (e.g., Beck et al. 2002).

Among the factors that cause fire behavior to vary within a boreal fire, the most important and perhaps the least well understood is the interaction between a spreading fire and local atmospheric conditions (Jenkins et al. 2001). It is common for boreal wildfires to strongly modify existing wind patterns or even produce their own windstorms (McRae and Flannigan 1990). As Forthofer and Goodrick (2011) noted in their review, there are many variations of these intra-fire whirlwinds, and they all cause sudden wind shears that drastically change fire intensity, spread rates, and temperatures, and that also produce spotting behavior. These within-fire whirlwinds are products of the convective forces generated by the intense heat of the burn; some even consider such whirlwinds to be a major form of fire spread during large crown fires (e.g., Haines 1982). Winds generated by wildfires occur in two general forms: whirlwinds that rotate around a horizontal axis at the sides of the convection column created by a spreading fire, referred to as "horizontal roll vortices", and whirlwinds that are created by the rising convection column and that rotate around a vertical axis, referred to as "vertical vortices" (Haines 1982, McRae and Flannigan 1990).

All of these weather factors produce complex and unpredictable post-fire patterns. In particular, horizontal roll vortices not only produce wind shears, but also

FORMATION OF WILDFIRE RESIDUALS

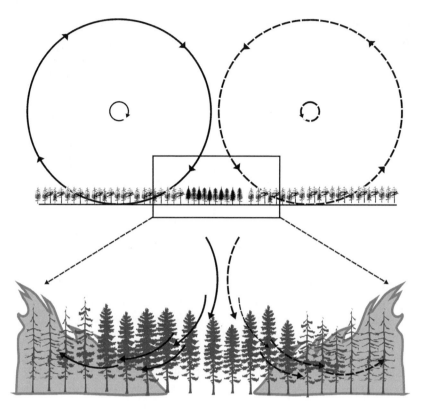

Figure 2.15   Internal whirlwinds created by fire can create unpredictable patterns of fire behavior. For example, horizontal roll vortices (top) produce long strips of unburned areas where the fire was driven away by the downward and outward movement of the wind. (Haines 1982. Reproduced with permission of American Meteorological Society.)

create unburned areas within a fast-spreading crown fire (Figure 2.15), which appear as long, narrow strips in the post-fire landscape (Haines 1982).

Process of residual formation

Wildfires transfer heat energy through three main sources: flames from combusting surface or above-surface fuels, buoyant plumes consisting of low-density heated air that rises from the flames, and smoldering combustion of surface or subsurface fuel (Dickinson and Johnson 2001, Miyanishi 2001). These processes transfer heat to different parts of the vegetation: flames and plumes transfer heat into the tree

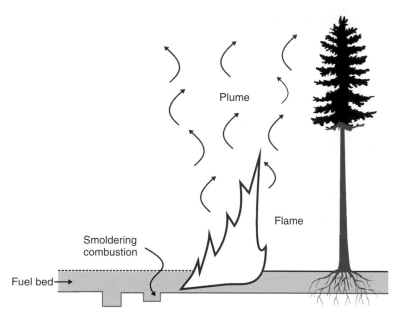

Figure 2.16   As fires burn, heat energy is transferred to vegetation by means of smoldering combustion that burns roots; flames that burn ground vegetation, tree stems, and the lower canopy; and plumes that burn the tree canopy. (Dickinson and Johnson 2001. Reproduced with permission of Elsevier.)

canopies, flames transfer heat to tree trunks and ground-level vegetation, and smoldering transfers heat to roots (Figure 2.16).

From this heat transfer, four outcomes for vegetation are possible. The vegetation can:

- escape damage by being sufficiently distant from the source of heat;
- receive transferred heat but survive by means of various adaptive mechanisms;
- die due to tissue or cell death caused by heating but remain structurally intact;
- burn almost completely and be reduced to charcoal and ash.

These eventualities occur through the interaction of the proximal fire intensity with the morphological characteristics of a species, and also with the physiological condition (e.g., moisture content) or phenological characteristics (e.g., in-leaf versus leaf-free) of an individual member of a species. Rarely will one fire produce a single, homogeneous outcome; in almost all boreal wildfires, the post-burn

FORMATION OF WILDFIRE RESIDUALS

patterns will include a combination of all of these outcomes. The following discussion focuses on variability in fire intensity, on the various tolerance mechanisms of plants, and on the consequent variability in vegetation responses to wildfire – all of which determine the formation of residuals in a boreal wildfire footprint.

Variability in fire intensity

The net result of heterogeneous fire behavior is a range of fire intensities within a wildfire footprint, and all classes of boreal fire types (Table 2.4) will be present in one or more parts of a fire. Given sufficient knowledge of fuel patterns, water bodies, and topography within a wildfire, it would be logical to expect that the windward side of ridges, drier sites, and areas with C-1 and C-4 fuel types would burn with higher intensity, whereas the leeward side of ridges and water bodies, moister sites, and areas with a D-1 fuel type would burn with a lower intensity. As a result of these differences, one would expect the heterogeneity of fire intensity to reflect pre-burn compositional and topographic patterns. However, variations in the spatial pattern of fire intensity may not be easily predictable simply through knowledge of pre-burn conditions and geo-environmental factors and of their expected interactions. Only a few general predictions are possible for fire intensity patterns, for example, one is a decreasing gradient from the interior of the wildfire footprint toward the fire perimeter.

This difficulty in generalizing and predicting fire intensity arises because the vagaries of fire weather and fire whirlwind patterns can easily confound or override the effects of relatively invariant factors such as fuel characteristics, fire barriers, and topography on a spreading fire. Furthermore, other within-fire processes, such as spotting and local extinguishment, add to the heterogeneity of fire intensity within a wildfire footprint. For example, spotting produces multiple ignition points and creates multiple fires with different frontal intensities (Figure 2.17). Depending on the wind direction and speed, these spot fires may or may not merge with the main fire. In addition, local extinguishment and smoldering can occur due to interactions among precipitation, wind, fuel type, and fire barriers. It has long been noted that atmospheric factors, including both the extreme fire weather that leads up to a fire and the intra-fire whirlwinds that develop, can override the effects of fuel patterns, water bodies, and topography on the heterogeneity of fire behavior and can thus affect post-fire vegetation patterns (Rowe and Scotter 1973, Haines 1982, Johnson 1992, Jenkins *et al.* 2001). This is especially true in the boreal forest, which has relatively flat, rolling terrain, where topography will have less influence on within-fire micro-weather patterns than it would, for example, in mountainous terrain. This overriding of effects makes it difficult to estimate spatial patterns of fire intensity solely on the basis of parameters known before the fire, such as fuel and topography. Therefore, almost all knowledge about fire intensity patterns within wildfires is derived from post-fire forensic analyses of the wildfire footprint

(a)

(b)

FORMATION OF WILDFIRE RESIDUALS

Figure 2.17   (a) Spot fires along the right flank of a 50 000 ha boreal wildfire evident from a high resolution satellite image and (b) an aerial view of the spotting pattern on the left front flank of a 40 000 ha boreal wildfire. These spot fires were ignited by firebrands carried ahead of the growing fire by high winds, and were able to pass over fire barriers such as water bodies.

using surrogates for intensity such as the crown scorch height (e.g., van Wagner 1973), although some limited research has been performed using a network of instruments during experimental burns.

Another consequence of the heterogeneity of fire behavior is the variability in temperature profiles (i.e., the peak temperatures reached at different heights or depths) within the forest canopy and the soil, and the residence time (i.e., the duration at a given location) of those peak temperatures (Whelan 1995). For example, at low wind speeds, the flames from a surface fire rise straight up and move slowly horizontally, and this pattern causes higher peak temperatures and longer residence times in the tree canopies. In contrast, when wind speeds increase, flames travel faster, so that the temperatures in their plumes are lower and the residence times are shorter (Mercer and Weber 2001). The vertical distribution of the fuels also affects the aboveground temperature profile. The presence of ladder fuels in the form of lower branches of tree canopies and shrubs in the undergrowth help the flames reach higher and increase peak temperatures and residence times in tree canopies (Whelan 1995). Also important is the belowground temperature profile produced by the fire. Because mineral soil is a good insulator, the peak temperatures reached 2–3 cm below the surface are relatively low; however, the magnitude of the surface and subsurface heating during a fire is affected by the fuel load, its moisture content, and how deeply the fire burns into the fuel bed (Whelan 1995). This is especially true with smoldering combustion, during which the surface and subsurface fuel may continue to burn for days or months after the fire front has passed and the surface fire has been extinguished (Miyanishi 2001).

Plant tolerance of heat damage

The critical temperature that changes the chemical composition in cells and tissues enough to cause tissue death appears to be 60°C, and cell death results after exposure to such lethal temperatures (van Wagner 1973, Whelan 1995, Gutsell and Johnson 1996). At higher temperatures, tissue and cell death occur at exponentially increasing rates, and the residence time required to reach lethal internal temperatures decreases (Dickinson and Johnson 2001). Damage to meristematic tissue can occur at three locations in a tree: in the trunk, the crown, and the roots. Boreal forest species that have evolved under a regime of frequent fire disturbances possess various adaptive mechanisms to avoid such tissue damage, or to recover *in situ* from the damage (Rowe and Scotter 1973, Rowe 1983). Such mechanisms allow individual plants to protect their meristematic tissue from cell death caused by heat transfer from flames on the trunk, plumes in the crown, or smoldering fires at the roots.

The likelihood of cambial cell death is higher at lower levels on the tree's trunk, near the base, where the flame temperatures are higher and the residence times are longer. As reviewed by Johnson (1992) and Dickinson and Johnson (2001), this is where the bark is also the thickest, although differences exist among species (Dickinson and Johnson 2004). Typically, within a particular species, larger diameter stems have thicker bark, and the bark is thicker at the base of older trees than of younger ones. Such thick bark on the trunk insulates vascular cambial cells against heat, and increases the residence time required for the flame to create lethal internal temperatures (Dickinson and Johnson 2001). Differential heating around tree trunks can occur with variations in wind flow characteristics, especially on the leeward sides of trees, which may be exposed to fast-moving flames for a shorter duration than the windward side. However, the wind eddies created around tree stems could cause flames to be higher and their residence times longer on the leeward side (Dickinson and Johnson 2001). Girdling occurs when cambial cells are killed around the entire circumference of the tree trunk, and the tree dies. However, when only some of the cambial cells are killed, fire scars form, and the surviving cambium grows around the wound to restore connectivity of tissues above and below the scar; in this case, the tree has a chance of surviving after the fire.

The crowns of boreal trees are typically killed when heated to 60 °C for 1 min by the fire's plume (van Wagner 1973). As mentioned, plumes consist of low-density heated air that rises above the flames; the air temperature and wind turbulence decrease with increasing height. Therefore, the key factors in transferring heat to canopies are the plume's temperature and its residence time in the crown; these are controlled by the height of the flame, the fire's horizontal velocity, and the presence and magnitude of vertical wind vortices. The key adaptive mechanism by which trees avoid the death of foliage is to develop a wider gap between the canopy and the ground and to undergo self-pruning of branches that eliminates ladder fuels and prevents flames from climbing directly into the canopy. The higher portions of the canopy are then subject only to the convective heat from the plume of surface fires. Variation in these characteristics causes boreal tree species to exhibit differences in the critical residence times of the peak temperature (Johnson 1992). Beyond the first-order (immediate) mortality caused by the destruction of all apical meristems on some trees (i.e., complete crown scorching), second-order or delayed mortality can occur due to partial death of crowns and apical meristems, and subsequent physiological changes (Michaletz and Johnson 2008).

Similar adaptive mechanisms, including distancing belowground meristem cells from direct heat to provide insulation, help some species recover from a fire. Since the soil is a good insulator, plants with a deeper root system, with latent buds in the stem base and roots, and plants with rhizomes can recover after fire through rapid sprouting even if the aboveground structure is scorched (Whelan 1995). However, the death of root tissue can occur from combustion of the soil organic layer by the moving fire, by smoldering combustion, or by a combination of both mechanisms. As is true for fire intensity and fire behavior, knowledge of boreal

FORMATION OF WILDFIRE RESIDUALS

tree mortality and survival during wildfires is based on post-fire empirical studies and correlations, and there is much room to improve the understanding of the associated physical and physiological mechanisms (Michaletz and Johnson 2007, Butler and Dickinson 2010).

Variability in fire severity

Fire severity is not a direct property of the fire, but instead represents the net result of interactions between fire intensity (ensuing peak temperatures and residence times) and the vegetation, both live and dead (Keeley 2009). Definitions of fire severity vary somewhat among the Canadian (CIIFFC 2003), American (NWCG 2012), and European (Stacey 2012) sources, though the overall meaning is similar. Although fire severity is a continuum, it can be grouped into classes related to the formation of wildfire residuals. These classes are high, medium, and low severity, and unburned (Table 2.5). These categories parallel the boreal fire regime classes of van Wagner (1983) and the fire severity classes of Keeley (2009). Because the fire intensity and the vegetation both vary spatially, all possible classes of fire severity may be evident within a boreal wildfire footprint.

Where fire severity is high, all plants are killed by direct combustion or by the death of meristematic cells. Active or passive crown fires at the fire front are the most probable cause for the destruction of tree canopies and all understory plants as well as the organic matter at the soil surface (Figure 2.18). In fact, over 80% of total biomass consumed in fires in boreal Canada consists of surface fuel (Zoltai et al. 1998, de Groot et al. 2009). Mineral soil is typically laid bare, and occasional heaps of ash can be found. Tree root systems are exposed by the consumption of organic matter, and root tissue death is likely. All trees are killed by the death of a combination of foliage, cambium, and roots. However, complete combustion or charring of most tree stems is not likely because of their high moisture content and the low conductivity of wood (Johnson 1992). Some researchers have linked fire intensity values directly to this severity class. For example, van Wagner (1983) considered that a fire intensity greater than 1500 kW m^{-1} would result in the survival of few trees within a boreal wildfire. Others have found similar fire intensity thresholds for the mortality of boreal trees (e.g., 2000 kW m^{-1}; Perera et al. 2009), beyond which no canopy trees survive immediately after the fire.

Medium fire severity results in a mixture of outcomes, with much spatial variability in the degree of mortality, especially in the tree crowns. With only sporadic crowning of fire in individual stems, much of the heat damage to trees may result from the flames produced by an intense surface fire, most likely at the fire front. Consequently, there will be a mixture of trees with burned crowns, partially green crowns, and completely green crowns (Perera et al. 2009). Destruction of the forest floor material is also variable; some understory plants may

Table 2.5 Generalized fire severity classes in wildfires, their characteristics immediately after the fire, and their effects on the formation of residuals.

Fire severity class and associated characteristics	Fire severity class (Keeley 2009)	Fire regime class (van Wagner 1983)
High severity		
Canopy trees are killed and foliage is consumed	Deep burning or crown fire	Class 3
Tree death is evident immediately after the fire		
All understory plants – shrubs, herbs, mosses, and lichens – are killed		
Litter and soil organic layers are completely consumed		
White ash deposits and charred organic matter are found to a depth of several centimeters		
Medium severity		
Some trees are killed, but not over large areas	Moderate or severe surface burn	Class 2
Some trees exhibit partial canopy death, but not all foliage is consumed		
Some trees have burn scars and scorching of the stem, but retain green foliage		
Most understory plants – shrubs, herbs, mosses, and lichens – are killed, but some survive	Light	
Litter and the soil organic layer are mostly consumed		
Charred organic matter is found to a depth of a few millimeters		
Low severity		
Canopy trees exhibit no scarring or scorching	Scorched	Class 1
Some understory plants are killed and scorched, but most trees survive with minimal to no damage		
Some litter exhibits only scorching as evidence of the fire		
Unburned		
No evidence of fire	Unburned	Not classified

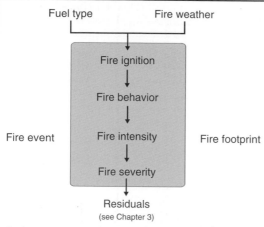

A boreal wildfire is typically started by a lightning strike that creates a point-source ignition. Dry, warm weather predisposes the living and dead forest biomass that constitutes the fuel type to burn by reducing its moisture content. Continuing dry weather coupled with windy conditions enables an ignited fire to grow, and as it expands, it releases thermal energy at high rates. During its spread, the fire front moves at variable speeds, consuming fuel on the forest floor at different rates, in response to the spatial patterns of fuel, topography, and prevailing weather. Collectively this reaction of the fire to fuel, topography, and weather is termed *fire behavior*.

As the behavior of a fire changes, the rate at which it releases thermal energy, that is, its *fire intensity*, also changes spatially and temporally. When fire intensity is high, the fire may inflict a high degree of damage on the vegetation, and when low, it may inflict little damage. The result of the interaction between a spreading fire and the live vegetation in its path is the *fire severity*, which indicates the degree of damage inflicted on the vegetation, including mortality. The post-fire remnant vegetation, whether dead or alive, represents the *residuals*.

The processes that occur from the ignition to complete extinguishment of a wildfire constitute a *fire event*, a term that identifies the temporal boundary of a wildfire. The signature of a fire event that creates a spatial boundary encompassing all fire processes is the *fire footprint*. Wildfire footprints are a rich source of forensic information, and have been the focal point of the studies that have provided much of the empirical knowledge of fire behavior, fire intensity, fire severity, and the formation of residuals.

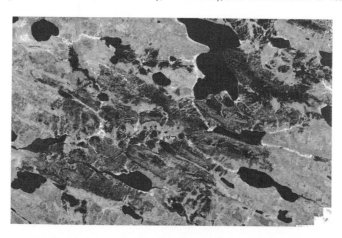

(a)

(b)

FORMATION OF WILDFIRE RESIDUALS

Figure 2.18   Where fire severity is high, the fire (a) kills all living vegetation and (b) burns the surface litter and organic matter into char and ash, exposing tree root systems.

(a)

(b)

FORMATION OF WILDFIRE RESIDUALS

Figure 2.19   Where fire severity is medium, (a) stems can have evidence of scorching and charring but retain green foliage, and (b) a mixture of burned and unburned stems is evident.

survive, as would patches of litter and organic matter (Figure 2.19). van Wagner (1983) indicated that a fire intensity that would cause this severity of damage was between 300 kW m^{-1} and 1500 kW m^{-1}. Under conditions with favorable fire weather indices and fuel types, medium severity could also result from a backfire. Tree mortality in the medium severity class, however, is time-dependent. Delayed mortality of trees that appear to have survived the fire based solely on the greenness of the foliage is rapid. The mortality rate can be as high as 40% after 1 year, and nearly 65% after 2 years (Perera *et al.* 2011), and there is evidence that post-fire tree mortality may continue for a decade in areas of medium severity (Angers *et al.* 2011). Explanations proposed for delayed tree mortality vary, ranging from simple variables (e.g., species differences; Angers *et al.* 2011) to interactions among variables (e.g., local fire intensity and initial mortality rates; Perera *et al.* 2011). However, it may not be possible to isolate the specific causal factor or factors in a given boreal wildfire because of the complexity of multi-scale interactions among causal factors and their high spatial variability (Dalziel and Perera 2009).

Low-severity fires are characterized by the near-absence or absence of scarred and scorched canopy trees. Little of the forest floor material is consumed, except for scattered patches of organic matter consumed by shallow, smoldering fires (Figure 2.20). The limited damage to understory plants is caused by the small flames of a creeping fire front or of a low-intensity backfire. Areas with low severity are also possible when local extinguishment occurs due to weather conditions or excessively wet sites, which may reduce the rate of fire spread as well as the intensity. Low fire severity is common along fire perimeters before fires are gradually extinguished. van Wagner (1983) hypothesized that low severity would result when the fire intensity is less than 300 kW m^{-1}.

The last fire severity class is characterized by areas with no evidence of fire. These appear as undisturbed islands in a matrix of disturbed areas within the wildfire footprint, and remain completely representative of the pre-burn vegetation at these locations. These areas have avoided the fire and are therefore called fire "skips". The existence of such unburned areas even within the largest and most intense boreal wildfires has drawn the attention of fire ecologists for decades (e.g., Lutz 1956), and many hypotheses have been proposed to explain their presence. These range from local extinguishment of fire fronts due to site or fuel conditions and fuel barriers (e.g., Scotter 1972) to changes in fire behavior due to external wind patterns, internal wind patterns, and spotting (e.g., Simard *et al.* 1983). Spotting may create unburned areas between the spot fire and the main body of the fire when the new ignition creates a low-intensity fire that consumes the fuel between its position and the main fire, creating a barrier to the spread of the main fire. In some instances, the unusual pattern of the unburned areas may provide clues to their formation, as is the case with the long, narrow strips of unburned forest that are formed by horizontal roll vortices (Figure 2.15; Haines 1982, Arsenault 2001).

FORMATION OF WILDFIRE RESIDUALS

FORMATION OF WILDFIRE RESIDUALS

(a)

(b)

Figure 2.20   Where fire severity is low, (a) patches of the understory and organic matter may be burned, but (b) the tree canopy and some of the understory vegetation remain unburned.

Figure 2.21   A typical boreal wildfire contains areas varying in fire severity from completely burned (with no living trees; dark) to partially burned (grey), to unburned (with no evidence of fire; light).

All of these causal factors are plausible, and it may not be possible to propose a single factor that explains the survival of these unburned areas, even within the same wildfire. A more detailed discussion of the characteristics and spatial patterns of these unburned areas is provided in Chapter 3.

All of the fire severity classes may be interspersed within a boreal wildfire footprint as a result of the heterogeneity of fire behavior. However, it is not uncommon for entire wildfires to be arbitrarily classified as a high-, medium-, or low-severity fire based on the proportion of the total area in each severity class, when in fact most boreal wildfires are of mixed severity, and contain all classes of fire severity (Figure 2.21).

Fire severity patterns also occur at finer spatial scales, nested within areas of a wildfire footprint that exhibit mixed fire severity. For example, patches with low fire severity, unburned islands, and (rarely) individual trees with unburned crowns may be found within areas of medium severity (Figure 2.22). This multi-scale mosaic of fire severity classes and the resulting juxtaposition of wildfire residuals make boreal wildfire footprints highly diverse, and initiate the many ecological flows and processes that are discussed in Chapter 4.

FORMATION OF WILDFIRE RESIDUALS

FORMATION OF WILDFIRE RESIDUALS

(a)

(b)

Figure 2.22   Fire severity patterns occur at many scales. For example, (a) an unburned area of shrubs and moss was found within a burned patch, surrounded by unburned vegetation, and (b) a single burned stem was surrounded by living vegetation in a lowland area.

Summary

The physics, chemistry, and biology of the various processes that govern wildfire behavior have long been studied, and this body of scientific knowledge continues to expand. This knowledge explains how individual components of the forest ecosystem, such as forest communities, topography, and weather, influence fire behavior and how the vegetation responds to intense heating during a wildfire. Many attempts have also been made to bring these individual components together to explain fire behavior and its impact on forest ecosystems, and thus the resulting spatial patterns of fire severity. To date, such efforts have been mostly empirical, relying on forensic investigations of post-fire patterns, and syntheses of mechanistic processes are scarce. For example, it is typical to explain the heterogeneity of boreal wildfire severity as a function of variability in the vegetation, geo-environment, season, and weather conditions, through consideration of a single factor at a time.

However, empirical explanations derived from past fires may not be reliable spatial predictors of fire severity, and thus of residual formation. The spatial manifestation of fire severity and residual formation cannot be predicted solely

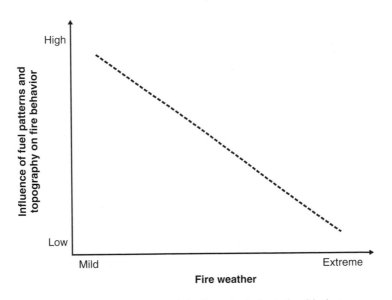

FORMATION OF WILDFIRE RESIDUALS

Figure 2.23 A simplified illustration of the hypothetical relationship between weather, topography, and fuel conditions, and the resulting effect on fire behavior and the formation of wildfire residuals. The effects of fuel and topography on fire severity and on wildfire residuals is most pronounced when fire weather is mild. Note, the slope is depicted as linear only for illustrative purposes.

from pre-burn knowledge of a single individual causal factor, or even from the sum of many causal factors. This is because the causal factors do not operate independently; rather, they are autocorrelated and produce an array of multi-factor interactions. Moreover, their influence appears to be hierarchically nested at different scales. For example, extremely dry and warm pre-fire weather, together with high winds during the fire, can override the effects of within-fire fuel barriers, while under relatively wet and cool weather, topographic factors and fuel patterns may dictate the post-fire heterogeneity of fire severity (Figure 2.23).

Furthermore, some explanatory variables change and others become significant during the fire (Figure 2.24). These factors alter the spatial patterns of fire behavior and fire severity in ways that are not predictable from empirical relationships. For example, spot fires that arise due to high winds can create multiple points of ignition that will create a discontinuity in fire severity that defies the patterns predicted from expectations of linear fire spread for a given set of conditions based on weather, fuel, and topography. As is the case for the internal cessation and initiation of fire spread, sudden conflagrations and positive feedback loops, such as internal wind vortices, can override the effects of other factors. Thus, spatial patterns of fire severity in boreal wildfires, which result in patterns of residual formation, appear to be an emergent property that cannot be simply predicted *a priori.*

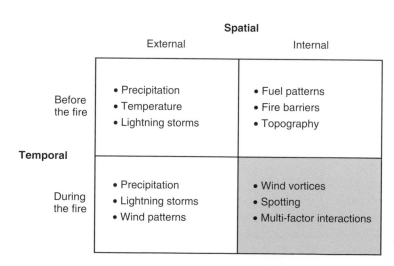

Figure 2.24 A conceptual depiction of the causal factors that affect fire behavior before and during the fire. Some factors originating externally exert an influence both before and during the fire (e.g., precipitation) and others (e.g., wind patterns) only during the fire. Other factors originate within the final wildfire footprint prior to the fire (e.g., fuel patterns) or during the fire (e.g., spotting). The shaded box indicates the causal factors that are emergent properties of the fire; these are the least well understood.

Consequently, we contend that present scientific knowledge is far from adequate to provide spatially and scale-explicit predictions of the formation, distribution, and characteristics of wildfire residuals. Although the known variables and processes can explain how wildfire residuals *may* form, it is not yet possible to predict with any degree of certainty *where, when,* and *how much* residuals will result from a boreal wildfire. This problem of knowledge gaps and uncertainty is not just an academic issue. As will be discussed in Chapter 5, many forest management agencies rely heavily on their knowledge of the spatial patterns of wildfire residuals to design forest management templates in the boreal region.

The solution to this problem may lie in how we study wildfire behavior and the resulting fire severity patterns. Instead of relying entirely on surveys of post-fire patterns and calculations of empirical correlations between these patterns and causal variables, and instead of limiting the study of mechanistic processes to a reductionist approach (i.e., one that concentrates narrowly on individual factors), it is necessary to seek a process-based understanding of the simultaneous interactions among many factors and the associated processes. The goal is to develop an integrated understanding of these processes, viewing each new wildfire as an opportunity to test and refine our hypotheses. This will be a slow, expensive, and challenging approach, but one that is needed if we hope to understand boreal wildfire severity patterns and how they affect the formation of wildfire residuals.

References

Alexander, M.E., Lawson, B.D., Stocks, B.J., and van Wagner, C.E. (1984) User Guide to the Canadian Forest Fire Behaviour Prediction System: Rate of Spread Relationships. Interim Edition. Environment Canada, Canadian Forest Service, Fire Danger Group, Chalk River, ON.

Angers, V.A., Gauthier, S., Drapeau, P., Jayen, K., and Bergeron, Y. (2011) Tree mortality and snag dynamics in North American boreal tree species after a wildfire: a long-term study. International Journal of Wildland Fire 20, 751–763.

Arseneault, D. (2001) Impact of fire behavior on post-fire forest development in a homogeneous boreal landscape. Canadian Journal of Forest Research 31, 1367–1374.

Beck, J.A., Alexander, M.E, Harvey, S.D., and Beaver, A.K. (2002) Forecasting diurnal variations in fire intensity to enhance wildland firefighter safety. International Journal of Wildland Fire 11, 173–182.

Butler, B.W. and Dickinson, M.B. (2010) Tree injury and mortality in fires: developing process-based models. Fire Ecology 6, 55–79.

Chandler, C., Cheney, P., Thomas, P., Trabaud, L., and Williams, D. (1983) Fire in Forestry: Volume II. Forest Fire Management. Wiley-Interscience, New York, NY.

CIIFFC (2003) The 2003 glossary of forest fire management terms. Canadian Interagency Forest Fire Centre, Winnipeg, MB. http://bcwildfire.ca/mediaroom/Backgrounders/2003_Fire_Glossary.pdf

Dalziel, B.D. and Perera, A.H. (2009) Tree mortality following boreal forest fires reveals scale-dependant interactions between community structure and fire intensity. Ecosystems 12, 973–981.

de Groot, W.J., Pritchard, J., and Lynham, T.J. (2009) Forest floor fuel consumption and carbon emissions in Canadian boreal forest fires. Canadian Journal of Forest Research 39, 367–382.

Dickinson, M.B. and Johnson, E.A. (2001) Fire effects on trees. In Johnson, E.A. and Miyanishi, K. (eds) Forest Fires: Behavior and Ecological Effects. Academic Press, San Diego, CA. pp. 477–525.

Dickinson, M.B. and Johnson, E.A. (2004) Temperature-dependent rate models of vascular cambium necrosis. Canadian Journal of Forest Research 34, 546–559.

Esseen, P.-A., Ehnström, B., Ericson, L., and Sjöberg, K. (1997) Boreal forests. Ecological Bulletins 46, 16–47.

Fendell, F.E. and Wolff, M.F. (2001) Wind-aided fire spread, forest fires, behavior and ecological effects. In Johnson, E.A. and Miyanishi, K. (eds) Forest Fires: Behavior and Ecological Effects. Academic Press, San Diego, CA. pp. 171–223.

Forthofer, J.M. and Goodrick, S.L. (2011) Review of vortices in wildland fire. Journal of Combustion doi:10.1155/2011/984363.

Goldammer, J.G. and Furyaev, V.V. (1996) Fire in Ecosystems of Boreal Eurasia. Kluwer Academic Publishers, Dordrecht, Forestry Sciences Vol. 48.

Gutsell, S.L. and Johnson, E.A. (1996) How fire scars are formed: coupling a disturbance process to its ecological effect. Canadian Journal of Forest Research 26, 166–174.

Haines, D.A. (1982) Horizontal roll vortices and crown fires. Journal of Applied Meteorology 21, 751–763.

Hirsch, K.G. (1996) Canadian Forest Fire Behavior Prediction (FBP) System: User's Guide. Natural Resources Canada, Canadian Forest Service, Northern Forestry Centre, Edmonton, AB. Special Report 7.

Hytteborn, H., Maslov, A.A., Nazimova, D.J., and Rysin, L.R. (2005) Boreal forests of Eurasia. In Andersson, F. (ed.) Coniferous Forests. Ecosystems of the World 6. Elsevier, Amsterdam. pp. 23–99.

Jenkins, M.A., Clark, T., and Coen, J. (2001) Coupling atmospheric and fire models. In Johnson, E.A. and Miyanishi, K. (eds) Forest Fires: Behavior and Ecological Effects. Academic Press, San Diego, CA. pp. 259–303.

Johnson, E.A. (1992) Fire and Vegetation Dynamics: Studies from the North American Boreal Forest. Cambridge University Press, Cambridge, UK.

Johnson, E.A. and Miyanishi, K. (eds) (2001) Forest Fires: Behavior and Ecological Effects. Academic Press, San Diego, CA.

Kafka, V., Gauthier, S., and Bergeron, Y. (2001) Fire impacts and crowning in the boreal forest: study of a large wildfire in western Quebec. International Journal of Wildland Fire 10, 119–127.

Keeley, J.E. (2009) Fire intensity, fire severity and burn severity: a brief review and suggested usage. International Journal of Wildland Fire 18, 116–126.

Larsen, J.A. (1980) The Boreal Ecosystem (Physiological Ecology). Academic Press, New York, NY.

Lutz, H.J. (1956) Ecological effects of forest fires in the interior of Alaska. USDA Forest Service, Alaska Forest Research Center, Juneau, AK. Technical Bulletin 1133.

McRae, D.J. and Flannigan, M.D. (1990) Development of large vortices on prescribed fires. Canadian Journal of Forest Research 20, 1878–1887.

Mercer, G.N. and Weber, R.O. (2001) Fire plumes. In Johnson, E.A. and Miyanishi, K. (eds) Forest Fires: Behavior and Ecological Effects. Academic Press, San Diego, CA. pp. 225–257.

Meyn, A., White, P.S., Buhk, C., and Jentsch, A. (2007) Environmental drivers of large, infrequent wildfires: the emerging conceptual model. Progress in Physical Geography 31, 287–312.

Michaletz, S.T. and Johnson, E.A. (2007) How forest fires kill trees: a review of the fundamental biophysical processes. Scandinavian Journal of Forest Research 22, 500–515.

Michaletz, S.T. and Johnson, E.A. (2008) A biophysical process model of tree mortality in surface fires. Canadian Journal of Forest Research 38, 2013–2029.

Miyanishi, K. (2001) Duff consumption. In Johnson, E.A. and Miyanishi, K. (ed., Forest Fires: Behavior and Ecological Effects. Academic Press, San Diego, CA. pp. 437–476.

NWCG (2012) Glossary of wildland fire terminology. National Wildfire Coordinating Group, National Interagency Fire Center, Boise, ID. Available from http://www.nwcg.gov/pms/pubs/glossary/index.htm (accessed January 2013).

Perera, A.H., Dalziel, B.D., Buse, L.J., and Routledge, R.G. (2009) Spatial variability of stand-scale residuals in Ontario's boreal forest fires. Canadian Journal of Forest Research 39(5), 945–961.

Perera, A.H., Dalziel, B.D., Buse, L.J., Routledge, R.G., and Brienesse, M. (2011) What happens to tree residuals in boreal forest fires and what causes the changes? Ontario Ministry of Natural Resources, Ontario Forest Research Institute, Sault Ste. Marie, ON. Forest Research Report No. 174.

Rothermel, R.C. (1972) A mathematical model for predicting fire spread in wildland fuels. USDA Forest Service, Intermountain Forest and Range Experiment Station, Ogden, UT. Research Paper INT-116.

Rowe, J.S. (1983) Concepts of fire effects on plant individuals and species. In Wein, R.W. and MacLean, D.A. (eds) The Role of Fire in Northern Circumpolar Ecosystems. Scope 18. John Wiley & Sons, Ltd, New York, NY. pp. 135–154.

Rowe. J.S. and Scotter, G.W. (1973) Fire in the boreal forest. Quaternary Research 3(3), 444–464.

Scott, G.A.J. (1995) Canada's Vegetation: A World Perspective. McGill–Queen's University Press, Montreal, QC.

Scotter, G.W. (1972) Fire as an ecological factor in boreal forest ecosystems of Canada. In Fire in the Environment: Symposium Proceedings. USDA Forest Service, Intermountain Forest and Range Experiment Station, Ogden, UT. Report No. FS-276. pp. 15–25.

Shugart, H.H., Leemans, R. and Bonan, G.B. (eds) (1992) A Systems Analysis of the Global Boreal Forest. Cambridge University Press, Cambridge, UK.

Simard, A.J., Haines, D.A., Blank, R.W., and Frost, J.S. (1983) The Mack Lake Fire. USDA Forest Service, North Central Forest Experiment Station, St. Paul, MN. General Technical Report NC-83.

Stacey, R. (2012) European glossary for wildfires and forest fires. European Forest Fires Network, Interregional Cooperation Program INTERREG IVC. Available from http://www.fire.uni-freiburg.de/literature/EUFOFINET-Fire-Glossary.pdf (accessed January 2013).

Taylor, S.W., Pike, R.G., and Alexander, M.E. (1997) Field guide to the Canadian Forest Fire Behavior Prediction (FBP) system. Fire Management Network, Northern Forestry Centre, Canadian Forest Service, Edmonton, AB. Special Report 11.

van Wagner, C.E. (1969) A simple fire-growth model. Forestry Chronicle 45, 103–104.

van Wagner, C.E. (1973) Height of crown scorch in forest fires. Canadian Journal of Forest Research 3(3), 373–378.

van Wagner, C.E. (1977) Conditions for the start and spread of crown fire. Canadian Journal of Forest Research 7, 23–31.

van Wagner, C.E. (1983) Fire behavior in northern conifer forests and shrublands. In Wein, R.W. and MacLean, D.A. (eds) The Role of Fire in Northern Circumpolar Ecosystems. Scope 18. John Wiley & Sons, Ltd, New York, NY. pp. 65–80.

van Wagner, C.E. (1987) Development and structure of the Canadian Forest Fire Weather Index System. Canadian Forestry Service, Ottawa, ON. Forestry Technical Report 35.

van Wagner, C.E. and Pickett, T.L. (1985) Equations and FORTRAN program for the Canadian Forest Fire Weather Index System. Canadian Forest Service, Petawawa National Forestry Institute, Chalk River, ON. Forestry Technical Report 33.

Weber, M.G. and van Cleve, K. (2005) The boreal forests of North America. In Andersson, F. (ed.) Coniferous Forests. Ecosystems of the World 6. Elsevier, Amsterdam. pp. 101–130.

Wein, R.W. and MacLean, D.A. (1983) An overview of fire in northern ecosystems. In Wein, R.W. and McLean, D.A. (eds) The Role of Fire in Northern Circumpolar Ecosystems. Scope 18. John Wiley & Sons, Ltd, New York, NY. pp. 1–18.

Whelan, R.J. (1995) The Ecology of Fire. Cambridge Studies in Ecology, Cambridge University Press, Cambridge, UK.

Zoltai, S.C., Morrissey, L.A., Livingston, G.P., and de Groot, W.J. (1998) Effects of fire on carbon cycling in North American boreal peatlands. Environmental Reviews 6, 13–24.

FORMATION OF WILDFIRE RESIDUALS

3 Types of wildfire residuals and their extent

As we discussed in Chapter 2, the incredible variability in fire intensity that occurs as a fire travels across a boreal landscape, together with the heterogeneity of the species composition, results in a complex spatial mosaic of vegetation that is burned to various degrees. This mosaic includes areas that are burned completely, partially, or not at all, all of which contain a combination of living and dead remnants of the structure of the pre-fire system. This remnant structure is collectively referred to as "post-fire residuals" here termed wildfire residuals. Those who have investigated these residuals have assessed them in many ways. For example, residuals have been considered at different scales, variously described or defined, investigated for specific research interests, studied at different times after a fire, and reported using a variety of metrics.

Given this wide range of definitions of and approaches to wildfire residuals in the scientific literature, it is possible that both novice readers and experienced

Ecology of Wildfire Residuals in Boreal Forests, First Edition. Ajith H. Perera and Lisa J. Buse.
© 2014 Ajith H. Perera and Lisa J. Buse. Published 2014 by John Wiley & Sons, Ltd.
Companion Website: www.wiley.com/go/perera/wildfire

researchers may be confused by the terminology and how the reported findings depend on that terminology. Therefore, our first goal in this chapter is to arrive at robust and generic definitions for the various types of wildfire residuals in the context of how they were formed (Chapter 2) and their potential role in post-fire systems (Chapter 4). Secondly, we discuss the known extent of wildfire residuals in boreal forests; that is, how much of each type of residual remains after a typical boreal fire and how that post-fire structure changes over time.

We begin by describing what one typically sees following a boreal wildfire and how that scene changes over time, then we summarize how these residuals can be perceived conceptually, since this perception determines both how they are defined and how they are quantified. We also summarize the results from surveys of various types of residuals with respect to their extent after boreal wildfires and to the post-fire changes that occur. These summaries provide a basis for a discussion of the challenges and limitations associated with current descriptions and assessments of the extent and characteristics of wildfire residuals.

In an attempt to improve on existing descriptions, we propose a harmonized set of definitions for wildfire residuals that are based both on their mechanism of formation and on their subsequent ecological functions; these working definitions will be applied throughout the remainder of this book. We then identify gaps in current knowledge and limitations of the study methods that have been used to describe and define wildfire residuals and their extent. Based on this analysis, we provide suggestions for how the study of wildfire residuals could be approached so as to fill the knowledge gaps and to provide a more standardized body of knowledge that will facilitate comparisons among studies.

Types of wildfire residuals

As a prerequisite to understanding specific descriptions and definitions of residuals, we begin by providing a generic description of the aftermath of a boreal wildfire and its constituents, followed by an overview of how these constituents have been described in previous studies.

A post-wildfire scene

First, we provide a detailed visual impression of the major components of the typical spatial mosaic that is found within a boreal wildfire footprint. (As defined in Chapter 2, the fire *footprint* is the spatial signature of a single fire event that encompasses all fire processes, bounded by a common perimeter.) For this purpose, we broadly define fire residuals as remnants of the pre-fire ecosystem that have

Figure 3.1   Viewed from the air, a complex mosaic of burned (dark), partially burned (grey), and unburned (light) areas is apparent after a large boreal wildfire resulting in a variety of residual structure.

retained at least some of their structure; that is, they were not entirely reduced to charcoal, ash, or gases during the fire. This collection of residuals within a wildfire consists of living and dead components, arranged in a spatially complex mosaic (Figure 3.1). The proportions of living and dead vegetation, as well as the spatial arrangement of the residual components, differ widely among and within fires due to the inherent variability, both in fire behavior and in the other contributing factors, as discussed in Chapter 2. The resulting spatial mosaic consists of the following: fully burned areas where all living structures died as a result of the fire, found where the fire severity was moderate to high; partially burned areas where the fire passed through but did not completely kill the vegetation, found where the fire severity was low to moderate; and unburned areas that the fire missed and therefore left completely unaltered within the fire footprint.

After large wildfires, the most visible components are the completely burned areas. These areas contain residual vegetation that did not survive the fire but was not reduced to ash, resulting in a vast pool of dead, standing and fallen, trees (Figure 3.2). Such residual stems will vary in size and density, depending on the structure of the pre-fire stands. The degree to which these stems are burned also varies, depending on the fire behavior. Though mostly charred, the stems provide habitat and become an important component of many post-fire

(a)

(b)

Figure 3.2   Completely burned areas result in an accumulation of dead standing and fallen stems. (a) Completely burned areas inside the wildfire footprint in a typical boreal forest can vary widely, as shown by the high abundance of dead standing stems that were charred to various degrees. (b) In such areas, the understory vegetation is killed and the forest floor organic matter is consumed by the fire, exposing the mineral soil.

ecological processes. On rare occasions, tree canopies may be completely burned by a fast-moving canopy fire that leaves the understory vegetation and forest floor unburned. However, typically, the understory vegetation and forest floor are also burned, leaving charred residuals below the canopy. Even in these completely burned areas, a reservoir of vegetative propagules remains, both above and below the ground, in the form of serotinous cones in the canopies of trees such as jack pine (*Pinus banksiana*) and black spruce (*Picea mariana*) and the root system and meristematic tissues of some tree species and shrubs, such as poplar (*Populus* spp.), birch (*Betula* spp.), and alder (*Alnus* spp.). As we discuss in Chapter 4, these aerial and underground propagules play an important role in the recovery of

Figure 3.3 Partially burned areas may contain trees that are alive, with crowns that escaped the fire or that were partially scorched, while the forest floor and the understory burned completely.

TYPES OF WILDFIRE RESIDUALS AND THEIR EXTENT

a burned boreal forest ecosystem. Even fiercely burned areas within a boreal wildfire, which may appear as lifeless as a desert soon after the fire, will show signs of regrowth and greening within weeks.

Any wildfire footprint is likely to contain a large proportion of partially burned areas within its perimeter (the external boundary that demarcates a fire footprint). These areas have been subject to the spreading fire, which was most likely a surface fire of low to moderate intensity, but contain some live components of the pre-fire system. Partially burned areas may include trees with live tree crowns where only the understory burned or where the trees burned but the crowns were not completely consumed (Figure 3.3). In both cases, some proportion of the tree canopy cover remains alive for a period after the fire. Similar patterns may be present in the understory and forest floor vegetation as well, where burned and unburned vegetation may be spatially interspersed (Figure 3.4). Due to varying fire intensities and to local extinguishment of the fire front, some vegetation may survive, creating an assortment of live trees and dead stems, occurring either individually or in clusters. Similar patterns of dead and live vegetation mixtures may be seen in the understory as well as on the forest floor. Some of these residual structures, although

Figure 3.4   Partially burned mosaics also occur on the forest floor. In areas where the trees are completely burned, the understory may be a mix of burned (dark), partially burned (grey), and unburned (light) areas.

they remain alive immediately after the fire, may undergo delayed mortality, but the rest will persist, contributing to the post-fire recovery of the ecosystem.

Partially burned areas are ecologically interesting for several reasons, as we elaborate in Chapter 4. First, they usually occur at the boundaries between burned and unburned areas, where they act as an interphase for ecological flows between these two contrasting components of the post-fire ecosystem. Secondly, they are highly dynamic structurally, as this is where most of the post-fire mortality and windthrow of trees occur. Thirdly, because of the abundance of stressed and dying trees, partially burned areas attract the first swarms of pyrophilous insects after a fire.

Unfortunately, the partially burned areas of boreal wildfires often appear to have eluded the attention of ecologists. Due to methodological reasons or the lack of clear definitions, partially burned areas are frequently grouped with burned or unburned areas, depending on the study objectives. In fact, some researchers (e.g., Viereck and Dyrness 1979) have gone to the trouble of differentiating between lightly, partially, and heavily burned areas but did not evaluate the ecological effects on the partially burned areas, which they defined as having 51–94% of the ground surface blackened.

As discussed in Chapter 2, there are several reasons why some portions of the wildfire footprint remain unburned, having entirely escaped the spreading fire fronts. These unburned areas show no evidence of fire (Figure 3.5a, b). Because they were not subject to the fire, these areas contain intact replicas of the pre-fire forest ecosystem, including its biota. Such remnants of unburned vegetation may be seen in varying sizes and locations throughout wildfire footprints. Although also seen at finer scales nested within the burned areas, these may elude detection depending on the scale of observation. For example, small unburned areas of understory or forest floor may exist within a large area of completely burned trees due to variability in fire behavior (Figure 3.4). Based on their location, size, and spatial arrangement, these unburned areas play a vital role in the recovery of a burned ecosystem. As we elaborate in Chapter 4, they act as a refuge for mobile animal species during the fire, and subsequently act as a source of propagules that spread into the burned and partially burned areas. A special case of unburned areas are forested islands within the many lakes that are interspersed in boreal landscapes that may escape burning, some repeatedly (Figure 3.6).

Past descriptions of wildfire residuals

Even though wildfire residuals can be broadly described on the basis of the fire behavior and other factors that influenced their formation, they are addressed rather variably in the scientific literature. A review of the literature reveals that

TYPES OF WILDFIRE RESIDUALS AND THEIR EXTENT

(a)

(b)

Figure 3.5 Unburned areas (light), where both the tree cover and the understory vege-tation remain intact, represent remnants of the pre-fire ecosystem that persist within the wildfire footprint, surrounded by areas that were burned (dark) or partially burned (grey) (a); seen from the ground, these intact forest systems remain untouched by the fire (b).

Figure 3.6 Unburned areas within wildfires include some islands that escape fire. In this example, the large island on the right escaped while the adjacent islands and the mainland burned during boreal wildfire.

the majority of the descriptions and categories of wildfire residuals are based on their state (i.e., dead or alive) when they were studied and on their structural features, which can be vertical (e.g., upright or downed tree stems) or spatial (e.g., the distribution of single trees or clusters). Based on their structural features, four generic types of tree-related fire residuals have been recognized in the literature (Table 3.1). Dead trees are referred to as *snags* (if they remain upright) or *downed wood* (if they are lying on the ground), and live residuals are referred to as *live trees* (if they occur individually) or *residual patches* (if they occur in clusters).

Table 3.1 Structural attributes and terms used to describe the main types of boreal wildfire residuals.

Post-fire state	Structural attributes	Commonly used term
Dead	Upright stems, whether intact or broken	Snags
	Partially or completely uprooted stems, broken stems lying on the forest floor	Downed wood
Live	Upright stems, with the canopy intact and green, that occur individually	Live trees
	Upright stems, with the canopy intact and green, that occur in contiguous clusters	Residual patches

TYPES OF WILDFIRE RESIDUALS AND THEIR EXTENT

A variety of terms have been used to further describe these four generic types of residuals. The structure of dead residuals is universally described in terms of individual entities (i.e., stems or logs) rather than groups; that is, there is no mention of snag patches or snag areas in the literature. Standing dead trees or snags are referred to as residual snags (Awada *et al.* 2004), as standing dead wood (Bond-Lamberty *et al.* 2003), as standing dead trees (Ohman and Grigal 1979, Schulte and Niemi 1998), as dead trees with no green needles (Beverley and Martell 2003, Nappi *et al.* 2003), and as fire-killed trees (Saint-Germain *et al.* 2004). Only rarely are stems that were dead before the fire differentiated from those created by the fire, as was done, for example, by Hobson and Schieck (1999). Downed wood is also described in many ways, ranging from downed dead wood (Bond-Lamberty *et al.* 2003, Haeussler and Bergeron 2004) to coarse woody debris or coarse woody material and logs (Harper *et al.* 2004, 2005). Downed wood generally includes only fallen stems, but in some cases also includes branches (Hely *et al.* 2000) and stumps (e.g., Saint-Germain *et al.* 2004).

Structures that are alive at the time of the study are described both as individual entities (i.e., live trees) and as clusters (i.e., residual patches). Live tree residuals have been referred to as live stems (Imbeau *et al.* 1999), individual surviving trees (Lutz 1956), and remnant trees (Ohman and Grigal 1979); however, the trees that were affected by fire but survived are rarely differentiated from those that escaped fire altogether. In one case, the degree of "greenness" was defined from the proportion of brown foliage (Beverley and Martell 2003), which indicated that the trees were, indeed, affected by fire to some extent.

In other cases, wildfire residuals have been characterized by their ecological roles; for example, they have been described as habitat trees, seed trees, or stream buffers. The terms used to describe residuals can also be related to their perceived utility or value; for example, the term "residual habitats" was used to describe clusters of residual trees that included both burned trees and unburned trees (Madoui *et al.* 2010). There are cases where post-fire snags have been characterized by the degree of damage, burning, or decay, when the researchers were interested in specific values such as the suitability for use by nesting or foraging birds (Murphy and Lehnhausen 1998, Nappi *et al.* 2003, Nappi and Drapeau 2011), or the amount of carbon and nutrients in the snags (Bond-Lamberty *et al.* 2003). The various terms used in the scientific literature to describe wildfire residuals are summarized in Table 3.2.

On some occasions, detailed criteria have been used to describe residuals even more narrowly on the basis of specific metrics such as the species, stem diameter, height, and lean angle for individual stems of both live trees and snags, or the species composition, extent, and location in the case of contiguous clusters or patches of live trees. The dimensions used to describe a given type of residual vary widely within the categories of residuals referred to in the literature, but

Table 3.2 Descriptions of wildfire residuals reported in the scientific literature based on the common categories cited in Table 3.1, illustrating the variety of terms that have been used and the main focus of each study.

Category		Term/description	Study focus	Source
Dead	Standing	Burned snags	Salvage logging	Schmiegelow *et al.* (2006)
			Wildlife	Hebblewhite *et al.* (2009)
		Dead trees with no green needles	Woodpeckers	Nappi *et al.* (2003)
		Fire-created snags	Woodpeckers	Nappi and Drapeau (2011)
		Fire-killed snags	Salvage logging	Bradbury (2006)
		Fire-killed standing dead woody debris	Carbon	Manies *et al.* (2005)
		Fire-killed trees, stems, or timber	Beetles	Saint-Germain *et al.* (2004)
			Decay fungi/insects	Wikars (2001)
			Salvage logging	McIver and Starr (2001)
				Saint-Germain and Greene (2009)
			Wood quality/beetles	Richmond and Lejeune (1945)
				Gardiner (1957)
				Ross (1960)
			Woodpeckers/beetles	Nappi *et al.* (2010)
		Post-fire snags	Salvage logging	Schwab *et al.* (2006)
		Residual snags of fire origin	Succession	Awada *et al.* (2004)
		Self-supporting standing dead trees	Fire residuals	Ferguson and Elkie (2003)
		Snags	Disturbance effects	D'Amato *et al.* (2011)
			Fire residuals	Perera *et al.* (2008, 2011)
			Songbirds	Hobson and Schieck (1999)
				Imbeau *et al.* (1999)
				Simon *et al.* (2002)
			Succession	Harper *et al.* (2004, 2005)
			Salvage logging	Hutto (2006)
		Standing dead trees	Fire effects	Viereck and Dyrness (1979)
			Songbirds	Schulte and Niemi (1998)
			Succession/nutrients	Ohman and Grigal (1979)
				Le Goff and Sirois (2004)
		Standing dead tree snags	Fire residuals	Perera *et al.* (2009a)

(continued overleaf)

TYPES OF WILDFIRE RESIDUALS AND THEIR EXTENT

Table 3.2 (*continued*)

Category		Term/description	Study focus	Source
		Standing dead wood	Carbon	Bond-Lamberty *et al.* (2003)
			Wildlife	Fisher and Wilkinson (2005)
		Standing snags from fire-induced mortality	Carbon	Slaughter *et al.* (1998)
	Fallen	Downed dead wood	Carbon	Bond-Lamberty *et al.* (2003)
		Downed wood	Fire residuals	Perera *et al.* (2008, 2011)
		Downed woody debris	Carbon	Manies *et al.* (2005)
		Downed woody material	Wildlife	Fisher and Wilkinson (2005)
		Logs	Succession	Harper *et al.* (2004)
		Logs and stumps	Beetles	Saint-Germain *et al.* (2004)
		Wind-fallen dead trees	Succession	Haeussler and Bergeron (2004)
Live	Trees	Individual surviving trees	Fire effects	Lutz (1956)
		Live residuals	Songbirds	Stuart-Smith *et al.* (2002)
		Live residual trees	Wildlife	Fisher and Wilkinson (2005)
		Live trees	Fire effects	Viereck and Dyrness (1979)
			Songbirds	Imbeau *et al.* (1999)
			Succession	Harper *et al.* (2004)
		Live tree residuals	Fire residuals	Perera *et al.* (2008, 2011)
		Remnant trees	Succession/nutrients	Ohmann and Grigal (1979)
		Residual live trees	Fire residuals	Perera *et al.* (2009a)
		Residual trees	Songbirds	Schulte and Niemi (1998) Hobson and Schieck (1999) Morissette *et al.* (2002)
			Succession	Haeussler and Bergeron (2004)
		Residual unburned trees	Fire effects	Kafka *et al.* (2001)
		Scattered trees	Fire effects	Scotter (1972)
		Surviving trees	Songbirds	Apfelbaum and Haney (1981)

(*continued overleaf*)

Table 3.2 (*continued*)

Category	Term/description	Study focus	Source
Patches	Fire skips	Songbirds	Stuart-Smith *et al.* (2002)
		Beetles	Pohl *et al.* (2008)
	Islands	Vegetation	Larsen (1962)
	Island remnants	Fire	Andison (2012)
	Isolates (unburned patches within the fire)	Songbirds	Stuart-Smith *et al.* (2002)
	Partially burned inclusions	Plant diversity	Rees and Juday (2002)
	Patches of green trees	Woodpeckers/beetles	Nappi *et al.* (2010)
	Preserved islands	Fire regime	Bergeron *et al.* (2002)
	Residual group	Fire residuals	Dragotescu and Kneeshaw (2012)
	Residual habitats	Fire residuals	Madoui *et al.* (2010)
	Residual patches	Fire residuals	Perera *et al.* (2009b)
	Residual patches of live trees	Songbirds	Schieck and Hobson (2000)
	Residual patches of unburned forest	Beetles	Gandhi *et al.* (2001)
	Residual islands	Fire effects	Kafka *et al.* (2001)
	Stringers	Fire effects	Lutz (1956)
	Unburned forest remnants	Fire residuals	Delong and Kessler (2000)
	Unburned islands within a burned area	Fire residuals	Eberhardt and Woodard (1987)
	Unburned patches	Beetles	Saint-Germain *et al.* (2004)
			Pohl *et al.* (2008)
		Plant diversity	Nowak *et al.* (2002)
	Unburned residual stands	Succession	Greene and Johnson (2000)
	Upland strings	Fire effects	Scotter (1972)

these represent the *lower limits* of the dimensions; residuals that fall below these pre-defined limits are ignored and are not reported in a given study.

Snags are commonly defined using the diameter of the main stem at breast height (DBH), with the minimum size limits varying widely, from >1 (e.g., Bond-Lamberty *et al.* 2003) to >15 cm (e.g., Imbeau *et al.* 1999), but with >10 cm as the most common (e.g., Simon *et al.* 2002, Ferguson and Elkie 2003, Saint-Germain *et al.* 2004). Occasionally, stem heights are also used to define

a snag, with height limits ranging from >0.5 (e.g., Harper *et al.* 2004) to >4 m
(e.g., Awada *et al.* 2004). Minimum sizes included in the assessments of downed
wood range from >1 (e.g., Bond-Lamberty *et al.* 2003) to >10 cm in diameter
(e.g., Saint-Germain *et al.* 2004), but with >5 cm as the most common minimum (Hobson and Schieck 1999, Haeussler and Bergeron 2004, Harper *et al.*
2004). Typically, these diameters are measured at the midpoint of the length of
downed wood.

As is the case for snags, live trees are described by diameters with lower limits
ranging from ≥2.5 (e.g., Apfelbaum and Haney 1981) to >10 cm (e.g., Imbeau
et al. 1999, Perera *et al.* 2009a) and by heights with lower limits ranging from
≥1 (Stuart-Smith *et al.* 2002) to ≥5 m (Imbeau *et al.* 1999). On one occasion, the
percent canopy cover was used as the metric to describe residual trees (Morissette
et al. 2002). Residual patches are generally described based on a minimum area of
a contiguous cluster of trees. These limits range from 0.01 (e.g., Dragotescu and
Kneeshaw 2012) to 1 ha (e.g., Eberhardt and Woodard 1987, Kafka *et al.* 2001,
Carlson *et al.* 2011), with the latter the most common.

These differences in the minimum dimensions reported for wildfire residuals
included in various studies are quite extensive, as illustrated in Figure 3.7.
Beyond introducing a fundamental ambiguity about what residuals are, this

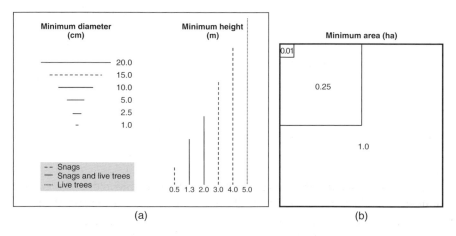

(a)   (b)

Figure 3.7 A schematic illustration of the wide range of dimensions used to describe the
major categories of post-fire residual structure as reported in the scientific literature for
(a) snags and live trees and (b) the area of residual patches. Sources for diameter (cm): 1.0
(Harper *et al.* 2005), 2.5 (Apfelbaum and Haney 1981), 5.0 (Harper *et al.* 2004), 10 (Perera
et al. 2009a), 15 and 20 (Imbeau *et al.* 1999). Sources for height (m): 0.5 (Harper *et al.*
2004), 1.3 (Haeussler and Bergeron 2004), 2.0 (Le Goff and Sirois 2004, Perera *et al.* 2009a),
3.0 (Ferguson and Elkie 2003), 4.0 (Awada *et al.* 2004), 5.0 (Imbeau *et al.* 1999). Sources for
area (ha): 0.1 (Dragotescu and Kneeshaw 2012), 0.25 (Perera *et al.* 2009b), 1.0 (Eberhardt
and Woodard 1987, Kafka *et al.* 2001, Carlson *et al.* 2011).

variability in terminology and dimensions also makes it very challenging, if not impossible, to compare different studies. For example, the extent of residuals is difficult to ascertain, given that the abundance of the residual structure differs dramatically depending on the definitions and minimum dimensions used to assess it.

There are also studies in which no specific metrics are mentioned but which provide qualitative descriptions of the vertical structure; for example, live trees are described only as "canopy" and "sub-canopy" trees (e.g., Hely *et al.* 2000). Furthermore, there are many studies that provide reports of abundance but do not provide quantitative definitions of wildfire residuals; this has been the case for snags (e.g., Hobson and Schieck 1999), live trees (e.g., Hobson and Schieck 1999), and residual patches (e.g., Bergeron *et al.* 2002, Madoui *et al.* 2010). Residual patches may include truly unburned areas as well as areas that were affected by fire, and separating the two can be an arduous task. For example, Eberhardt and Woodard (1987) acknowledged that assessing residual patches using aerial photographs at a 1-ha resolution did not let them distinguish whether areas that they called "unburned islands" were truly unaffected by fire because although the canopy appeared alive, some of the islands may have been "underburned" (i.e., only the understory and forest floor burned). Some authors have specified non-burned areas (e.g., Scotter 1972, Schieck and Hobson 2000) or unburned areas (e.g., Nowak *et al.* 2002) as "patches", whereas others were less specific but clearly referred to areas that were at least partially burned (Rees and Juday 2002). Others acknowledged outright that their so-called "residual patches" included both burned and partially burned categories. Some authors even specified the degree of burning within the residual patches included in their assessments (e.g., Andison 2012, Dragotescu and Kneeshaw 2012).

Abundance and extent of wildfire residuals

It is evident that descriptions in the scientific literature of the abundance and extent of the residuals after boreal wildfires must be considered in the context of how the authors defined residuals. This is because most assessments of wildfire residuals have been conducted to answer a specific question, and these different questions may require different definitions. As we noted in the previous section, residuals are typically assessed on the basis of their specific role in some aspect of post-fire ecology (e.g., insect invasions, woodpecker habitats, carbon sequestration, tree regeneration; Table 3.2). Consequently, these estimates of the extent of residuals are partial rather than holistic depictions; that is, the residuals have been examined without considering the patterns of fire severity throughout the fire, the complete range of types of residuals therein, and the characteristics of the fire footprint. However, given the number of studies of boreal wildfire residuals in the scientific literature, we provide a summary of the findings based on the main categories of standing residuals that we described earlier: snag residuals, live tree residuals, and residual patches. In addition, we present various suggestions that have been made in the literature about where residuals are likely to be found within a wildfire footprint.

Snag residuals

Based on the diverse qualitative and quantitative definitions of snag residuals, various metrics have been used to report their abundance after boreal wildfires. Most researchers have used snag density (the number of stems per hectare) and basal area (m^2 ha^{-1}) to report their abundance, but others have resorted to snag volume (e.g., Bond-Lamberty *et al.* 2003) or the number of stems per length of transect (e.g., Graf *et al.* 2000). Regardless of how snags were defined and measured, there have been no reports of a complete absence of snags after boreal wildfires, and given the number of studies, this provides evidence that at least some

Table 3.3   Some examples of estimates of snag abundance reported after wildfires in North American boreal forests.

Abundance		Dimensions used in assessment		Number of fires studied	Years after the fire	Study focus	Source
Density (stems ha^{-1})	Basal area (m^2 ha^{-1})	Diameter (cm)	Height (m)				
238–792	–	>10	≥2	11	0	Fire residuals	Perera *et al.* (2009a)
263	–	>10	>3	–	10–40	Fire residuals	Ferguson and Elkie (2003)
419	–	–	>4	–	<1-200	Tree regeneration	Awada *et al.* (2004)
600–2400	–	>1	>1	1	<50	Forest succession	Harper *et al.* (2005)
675	–	<10 10–14.9 15–19.9 ≥20	≥2	–	"Recent"	Songbirds	Imbeau *et al.* (1999)
754–2184	–	>2	–	–	11	Forest succession	Le Goff and Sirois (2004)
1710	–	>5	>0.5	2	3–4	Forest succession	Harper *et al.* (2004)
–	9.7–35.5	–	–	–	1	Songbirds	Hobson and Schieck (1999)
–	23	–	>1.3	4	3	Forest succession	Haeussler and Bergeron (2004)

–, Not reported.

snag residuals are always present. The range of snag densities found in wildfire footprints varies considerably (Table 3.3), from as few as 240 per hectare (Perera *et al.* 2009a) to as many as 2400 per hectare (Harper *et al.* 2005). Basal area estimates of snags have ranged from 9.7 to 35 m² ha⁻¹ (Hobson and Schieck 1999). Interestingly, given that a major determinant of the abundance of residual snags is the pre-burn tree density, no published studies have reported the post-fire snag abundance as a proportion of the pre-fire stand density, so it is unclear what proportion of the snags resulted from the fire. As can be seen from the examples in Table 3.3, the metrics that have been used to assess snags vary widely, and inconsistency in the sampling approaches makes it difficult to generalize from these reported abundance values.

As we discussed in Chapter 2, the frontal fire intensity of a boreal wildfire is a direct determinant of tree mortality, and hence of the production of snags. Indeed, some have used the variability of fire intensity to explain the variability of snag abundance within wildfire footprints. For example, Apfelbaum and Haney (1986) indicated that lower fire intensities cause less tree mortality and therefore result in fewer residual snags. However, other explanations have also been offered for the spatial distribution of snags. Miyanishi and Johnson (2002) implied that species and site differences existed, indicating that black spruce stands had more snags than mixed jack pine – black spruce stands; similarly, Bond-Lamberty *et al.* (2003) suggested that well-drained boreal black spruce sites have a higher abundance of snags than poorly drained sites. Other researchers (e.g., Apfelbaum and Haney 1986, Harper *et al.* 2005) have reported that spatial variability in site conditions can cause variability in the occurrence of residual snags, even though specific details of these associations were not provided. Stand age appears to be another important source of variability that determines the proportion of the pre-fire trees that become snags after a fire (Apfelbaum and Haney 1986). It is difficult to generalize such explanations because, in addition to the differences in the definitions and metrics that have been used to study snags, most of these investigations have not examined the whole suite of factors that potentially affect the conversion of living trees into snags, notably the fire behavior and its frontal intensity. This point is well exemplified in the next section, in which we discuss the explanations that have been proposed for the variability in the number of live residual trees – the living counterpart of snag residuals.

Live tree residuals

A broad range of abundance values have been documented for live tree residuals, mostly in terms of the number of stems per hectare. These include reports of no live trees (e.g., Viereck and Dyrness 1979, Imbeau *et al.* 1999, Stuart-Smith *et al.* 2002),

TYPES OF WILDFIRE RESIDUALS AND THEIR EXTENT

Table 3.4 Some examples of estimates of the abundance of live trees reported after wildfires in North American boreal forests.

Abundance		Dimensions used in assessment		Number of fires studied	Years after the fire	Study focus	Source
Density (stems ha^{-1})	Basal area (m^2 ha^{-1})	Diameter (cm)	Height (m)				
0	–	–	–	–	1	Songbirds	Hobson and Schieck (1999)
0	–	<10 10–14.9 15–19.9 ≥20	≥5	–	"Recent"	Songbirds	Imbeau et al. (1999)
0–10.4	–	–	>1	1	5	Songbirds	Stuart-Smith et al. (2002)
0–697[a]	–	All sizes	All heights	1	1	Fire effects	Viereck and Dyrness (1979)
78–316[b]	–	>10	≥2	11	0	Fire residuals	Perera et al. (2009a)
97	–	>5	–	1	2–3	Songbirds	Schulte and Niemi (1998)
120	–	>5	–	2	3–4	Forest succession	Harper et al. (2004)
–	0–12	–	>1.3	4	3	Forest succession	Haeussler and Bergeron (2004)
–	50.8	≥2.5	–	1	1	Songbirds	Apfelbaum and Haney (1981)

–, Not reported.
[a] 0 in heavily burned areas, 697 in lightly burned areas.
[b] After 2 years, an average of 13% of the live trees remained; the rest became snags or downed wood.

of <100 stems per hectare (e.g., Stuart-Smith et al. 2002, Perera et al. 2009a), of 100–500 stems per hectare (e.g., Schulte and Niemi 1998, Harper et al. 2004, Perera et al. 2009a), and of >500 stems per hectare (e.g., Viereck and Dyrness 1979). Less frequently, their abundance has been reported in terms of basal area per hectare, with values ranging from 0 (Haeussler and Bergeron 2004) to >50 m^2 ha^{-1} (Apfelbaum and Haney 1981). As with snag residuals, abundance values of live tree residuals in reports also depend on the study methods, objectives, and the definitions or dimensions used to describe them (Table 3.4). The timing of sampling after the fire also varies among studies, from immediately after the fire to 5 years later. As we discuss later in this chapter, studies conducted soon after a fire may greatly overestimate the abundance of live trees.

Fire intensity is often reported as an important factor in determining the occurrence of live tree residuals (Lutz 1956, van Wagner 1973, Methven et al. 1975, Ohman and Grigal 1979, Morissette et al. 2002, Hely et al. 2003, Haeussler and Bergeron 2004, Perera et al. 2009a). As we discussed in Chapter 2, the frontal fire intensity changes considerably within a wildfire event due to many factors, such as diurnal shifts in wind speed, heterogeneity in topography, and internal wind shear. Consequently, various locations inside a wildfire footprint may be subject to lower or higher frontal fire intensities, which will influence the probability of live tree occurrence at these locations (Methven *et al.* 1975, Morissette *et al.* 2002, Hely *et al.* 2003, Perera *et al.* 2009a). For example, it has generally been assumed that the occurrence of live tree residuals is high near fire perimeters because of lower fire intensities at the edges of a fire (e.g., Lutz 1956, Haeussler and Bergeron 2004). In fact, Harper *et al.* (2004) found that the density of live tree residuals decreases rapidly with increasing distance inward from the fire perimeter. Furthermore, Methven *et al.* (1975) reported a high occurrence of live tree residuals on the flanks of a fire and in areas where the fire spread at night, as this location and time period are typically associated with low frontal fire intensities. Many other suggestions have been offered in the scientific literature to explain variation in fire intensity, but the most common relates to the effects of local site conditions. Areas with a higher moisture regime, such as wetlands, lower slopes, and poorly drained areas such as black spruce bogs, are believed to have the highest occurrence of live tree residuals (Lutz 1956, Scotter 1972, Nordin and Grigal 1976, Viereck and Dyrness 1979, Apfelbaum and Haney 1981). Conversely, the occurrence of live tree residuals is considered to be least likely in well-drained areas with a drier moisture regime (Lutz 1956, Scotter 1972).

In addition to variability in fire intensity, differences in fire tolerance among species may determine the spatial patterns of live tree residuals (e.g., Scotter 1972, van Wagner 1973, Ohman and Grigal 1979). For example, mature balsam poplar (*Populus balsamifera*) is reported as being the most fire-resistant boreal tree species due to its bark thickness, followed by jack pine, black spruce, and trembling aspen (*Populus tremuloides*) (Lutz 1956, Scotter 1972). Reports of the survival of coniferous trees are mixed, but their survival rates are generally considered to be higher than those of deciduous species (Hobson and Schieck 1999), especially for large trees (Ohman and Grigal 1979, Hauessler and Bergeron 2004). Deciduous cover types, except balsam poplar, generally produce few or no live tree residuals. White birch (*Betula papyrifera*) does not readily survive because the bark of young trees is thin and the thicker, peeling bark of older trees is highly flammable (Lutz 1956, Scotter 1972). However, deciduous cover types such as aspen and birch may have live tree residuals despite their individual vulnerability to fire because these stands are generally less flammable and produce stems that lack ladder fuel. In contrast, spruce forests often have a fuel-rich understory and their crowns serve as ladder fuels, making them more vulnerable to fire (Epting and Verbyla 2005). Aspen has thick bark, but deep fissures in the bark increase heat conduction into the stem, making these trees less likely to survive for more than a year after a fire (Hely *et al.* 2003). Aspen residuals may occur at the edges

TYPES OF WILDFIRE RESIDUALS AND THEIR EXTENT

of a fire, especially if the stems are large, but will rarely be found in the fire's interior, where fire intensity is generally higher (Haeussler and Bergeron 2004). In general, smaller trees, with smaller diameters, shorter heights, and lower crown bases, have the lowest survival rate (Kafka *et al.* 2001, Beverley and Martell 2003, Hely *et al.* 2003). Although survival rates can vary among species and fire intensities, trees >10 cm in diameter are generally more likely to survive than smaller trees (Beverley and Martell 2003).

Given the paucity of abundance values in the scientific literature and the very different forest and fire conditions under which these studies are conducted – including pre-burn stand characteristics (e.g., cover type, age, size, density), site variables, fire behavior, time since the fire, and definitions of live tree residuals – it is difficult to compare abundance and extent of live tree residuals. Although the many *post hoc* hypotheses that have been proposed to explain the distribution of live tree residuals within wildfire footprints are plausible, our discussion in Chapter 2 showed that the survival of trees within a boreal wildfire is a product of the interaction between frontal fire intensity and the tolerance of a tree species, both of which are products of an array of physical and biological factors. Results from testing hypotheses about why some trees survive and where they occur within the footprint of boreal wildfires can only be generalized to other sites through a clear understanding of the causal factors and the mechanistic processes that govern fire behavior and that determine the fire tolerance of trees.

Residual patches

The extent of residual patches has been reported in several ways in the scientific literature, but mostly as the proportion of an area that did not burn (Table 3.5). This proportion ranges from 0% in some fires (e.g., Eberhardt and Woodard 1987, Perera *et al.* 2009b) to 22% in others (e.g., Madoui *et al.* 2010). The average size of the residual patches varies as well, and can be as large as 43 ha (Kafka *et al.* 2001), but most authors have reported average patch sizes of <10 ha (Eberhardt and Woodard 1987, Gluck and Rempel 1996, Perera *et al.* 2009b, Carlson *et al.* 2011). Although the distribution of residual patches varies significantly within a fire footprint, as a function of fire size, the occurrence of residual patches may be spatially autocorrelated. For example, in the 69 fires they studied, Eberhardt and Woodard (1987) found that the extent of residual patches decreased as the distance to neighboring residual patches increased (from 100 to 500 m). However, such trends must not be generalized because of the high variability in residual patch characteristics from one fire to another and the range of definitions used to identify residual patches (Table 3.5).

Table 3.5 Reported estimates of residual patch characteristics in North American boreal wildfires.

Patches include	Basis for estimating the proportion of residual patches	Proportion of fire area accounted for by residual patches (mean % of total)	Mean area of patches (ha)	Min. patch size (ha)	Number of fires studied	Years after the fire	Fire size (ha)	Source
Only unburned	Burnable area	0–11.74 (3.7)	0.98	0.25	17	2–7	57.7–4525.3	Perera et al. (2009b)
	Burned area	1.5–17 (5)	–	–	16	–	4509– 66 655	Bergeron et al. (2002)
	Burned area	2.6	43	1	1	0	49 070	Kafka et al. (2001)
	Total fire area	6.6 8.6	–	–	2	0	24 995 4641	Nappi et al. (2010)
	Total fire area	12.8	3	1	1	0	1200	Carlson et al. (2011)
Unburned and burned	Disturbed area	0–5.2	2.3–9.4	1	69	"Soon after fire"	21–17 770	Eberhardt and Woodard (1987)
	Total fire area	2–22 (10.4)	–	–	33	<25	2026– 51 882	Madoui et al. (2010)
	Burnable area	7.3–19.1	–	0.01	–	<10	136–7976	Dragotescu and Kneeshaw (2012)
	Burned area	11	–	–	24	0–7	28–15 909	Andison (2012)
Not specified	Burned area	21	9.6	–	1	9	18 415	Gluck and Rempel (1996)

–, Not reported.

TYPES OF WILDFIRE RESIDUALS AND THEIR EXTENT

Figure 3.8   Residual patches in wildfires show the spatial patterns of fire behavior. These linear residual patches are thought to have occurred due to the development of horizontal roll vortices within a large boreal wildfire. See Chapter 2 for a description of these vortices. (Photo: D. Schroeder. Reproduced with permission from FPInnovations.)

The occurrence of residual patches appears to be closely associated with spatial variations in fire behavior (Figure 3.8). For example, internal changes in fire behavior can lead to phenomena such as the development of horizontal roll vortices, thereby creating concentric narrow strips of residual patches that have been termed "streets" (Haines 1982). This effect is also evident from the finding that low nighttime fire intensities, which are typical in boreal wildfires, result in a high incidence of residual patches (Simard *et al.* 1983). Large fires have a higher probability of encountering varying fuel types and weather conditions. In fact, Kafka *et al.* (2001) reported that fires that burn for long periods may produce high variability in the occurrence of residual patches because the weather conditions – and therefore the fire behavior – are more variable. Similarly, Epting and Verbyla (2005) noted that the variation in the patterns of residual patches within fires may result from differences in fire behavior caused by changes in fuel moisture content. Changes in fire behavior due to neighborhood effects, including the proximity to less flammable fuel, can also cause variability in the occurrence of residual patches (Simard *et al.* 1983). For example, Epting and Verbyla (2005) showed that deciduous forest and shrub stands may act as landscape-scale barriers to the spread of a fire, likely because of their lower flammability and lack of crown ladder fuels. However, proximity to highly flammable fuel types can cause even inherently less-flammable cover types to form fewer residual patches (Simard *et al.* 1983).

In general, post-fire residual patches have been found to be associated with moist areas, such as wetlands and lowlands (Arseneault 2001, Nowak *et al.* 2002, Rees and Juday 2002). One common observation in the literature is that fires skip most wetlands and other moist areas, and leave unburned islands in their proximity (e.g., Arseneault 2001, Nowak *et al.* 2002, Rees and Juday 2002). Low-lying areas such as depressions (Buech *et al.* 1977) and swales (Nowak *et al.* 2002, Rees and Juday 2002) are also thought to act as fire barriers because they are moist and thus more difficult to burn; as a result, their presence increases the number or extent of residual patches within wildfires. This is also true for coniferous stands growing on organic deposits, which are less subject to crown fires and are therefore more likely to remain as residual patches (Kafka *et al.* 2001). Larger fires may have more internal variability in site conditions and present more opportunities for encountering fire breaks as they spread downwind, so large residual patches are more likely to occur (Eberhardt and Woodard 1987). There are also many examples in the literature of the presence of water bodies being used to explain the occurrence of residual patches. Post-fire residual patches are generally considered to be associated with open bodies of water, including both lakes and streams, and they are thought to be more common on the leeward side of lakes (Larsen 1962, Kafka *et al.* 2001). Some researchers consider pre-burn forest cover composition to be an important explanatory variable for the occurrence of residual patches, in that areas of lowland conifers are likely to contain a relatively high number and extent of residual patches (e.g., Kafka *et al.* 2001, Epting and Verbyla 2005).

Such effects of natural fire barriers within wildfire footprints are typically interpreted from post-fire evidence of the distribution of residual patches and their spatial association with various landscape characteristics; the effects are not inferred based on actual observations of fire behavior. Furthermore, there is a high degree of variability among fires in the spatial associations between residual patches and site conditions or bodies of water (Perera *et al.* 2009b). Therefore, we caution readers that such associations should be treated as hypotheses, not general trends.

As with published information about other categories of residuals, it is difficult to generalize the published knowledge of residual patches because the study methods, as well as the definitions of residual patches and their minimum sizes, vary among studies (Table 3.5). Furthermore, the proportions of the fire footprint occupied by residual patches have been estimated using different denominators. In such cases, the same residual extent would produce reports of different residual proportion depending on whether the basis for the estimate was the *total fire area* (the area bounded by the external fire perimeter), the *burnable* area (the area of all cover types that did burn and that could have burned, excluding bodies of water

TYPES OF WILDFIRE RESIDUALS AND THEIR EXTENT

and areas with no fuel, such as exposed bedrock), or the *burned* area (area burned within the external fire perimeter).

Changes in residuals after wildfires

The residuals left behind by forest fires, in the form of live tree patches, individual live trees, snags, and downed wood, are dynamic: their extent and spatial patterns change over time. The residuals may also shift among categories over time. For example, through delayed mortality, a small residual patch may become a scattered group of snags with intermittent live trees, or live trees may fall and become downed wood. These changes could occur rapidly, through an episodic event such as storm-related windthrow of all live tree residuals, or could develop gradually, as the trees age, decay, break, and fall. However, many reports of the extent of wildfire residuals have focused only on the period immediately after the fire and represent a sample at only a single point in time, even though wildfire residuals are transient, in both the short term and the long term. Although it may be onerous to track the temporal trends of residuals and identify the causal factors responsible for their changes, it is important to recognize that residuals change over time. In this section, we discuss the temporal changes in the major categories of residuals during the initial period of rapid change. We do not address slower, long-term changes in residuals that result from the breakdown of downed woody material and its decomposition.

Snag residuals

It is common for snag residuals to break and become downed wood after a fire (Figure 3.9). Most researchers agree that few snags fall within the first 3–5 years after a fire (Ohman and Grigal 1979, Viereck and Dyrness 1979, Apfelbaum and Haney 1981, Nalder and Wein 1999, Perera *et al.* 2011). Snags appear to fall and become downed wood mostly between 5 years and 15 years after a fire (Hobson and Schieck 1999, Awada *et al.* 2004); snag fall rates vary from 50 (Schaeffer and Pruitt 1991) to 100% (Nalder and Wein 1999) during this period.

Some researchers have observed that the post-fire fall of snags is related to the local fire severity. Both Angers *et al.* (2011) and Boulanger *et al.* (2011) reported that snag fall was lower in areas where the forest had burned more severely, and they attributed this trend to the low moisture content in the stems of heavily burned snags. Others have observed that the rates of snag fall differ among species and site conditions. For example, Lutz (1956) noted that high rates of windthrow of black spruce snags occurred 3 years after a fire on shallower soils due to root mortality during the fire. Similarly, in one study all aspen snags fell within 5 years

Figure 3.9 Over time, snag residuals break and fall to the ground, adding to the pool of downed wood. These changes continue to occur slowly during the first few decades after a fire.

after a fire (Nalder and Wein 1999). In another study, less than half of the jack pine and black spruce snags fell within the first 5 years, but few remained standing 10 years after the fire (Schaeffer and Pruitt 1991). A further study reported that after 10 years, 33% of the jack pine snags, 20% of the aspen snags, and 15% of the spruce snags remained standing (Angers *et al.* 2011). White spruce (*Picea glauca*) snags (Awada *et al.* 2004) and mixedwood snags (Apfelbaum and Haney 1986) fell within 15 years of a fire. In one simulation study, rates of snag fall were positively correlated with the post-fire snag density up to 10 years after a fire (Perera *et al.* 2011). Initially, this depletion of snags is countered by delayed tree mortality, which adds new snags to replace old ones that have fallen (Viereck and Dyrness 1979). Consequently, although snag residuals continue to fall, they may be seen within wildfire footprints for more than a decade after a fire (Figure 3.10).

Live tree residuals

Post-fire dynamics of live tree residuals are seldom reported. Documented studies include those by Hely *et al.* (2003), who monitored live trees within the first year after a fire, Perera *et al.* (2011) within 2 years, Viereck and Dyrness (1979) within

Figure 3.10   Snag residuals can be seen in old fire footprints for a long time after fire. In this case, jack pine snags are still evident 14 years after a boreal wildfire.

3 years, and Angers *et al.* (2011) for 10 years. All found that trees that appeared to be alive immediately after the fire undergo delayed mortality at high rates: 10–100% within 4 months (Hely *et al.* 2003), 40% after 1 year and 63% after 2 years (Perera *et al.* 2011), and 91% after 3 years (Viereck and Dyrness 1979). Angers *et al.* (2011) reported that there was a distinct pulse of tree mortality during the first 2 years after a fire and that mortality continued at a lower rate for up to a decade, even in low- and moderate-severity boreal fires. Although some have suggested that mortality rates differ among species, with the highest rates in aspen, followed by black spruce and jack pine (Hely *et al.* 2003), others consider local fire intensity to be the primary determinant of delayed mortality in live tree residuals (Viereck and Dyrness 1979, Perera *et al.* 2011). If the live trees do not become snags through delayed mortality, they may be uprooted or broken by the wind (Figure 3.11). We have observed abundant windthrow of live trees on exposed ridges after a fire and Apfelbaum and Haney (1981) noted that many live tree residuals were windthrown within a year after a fire, especially in and around bogs. From a study of four boreal wildfires, Perera *et al.* (2011) found that live tree residuals fell at a decreasing rate during the first 2 years after the fire (Figure 3.12). These rates were not related to the species or stem diameter, but were positively correlated with the fire intensity and negatively correlated with the initial stem density. The difficulties of relating live tree mortality to isolated

Figure 3.11   Live tree residuals that remain after boreal wildfires are frequently uprooted or broken during the first year after the fire.

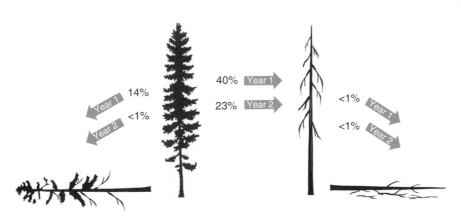

Figure 3.12   The fate of live tree residuals during the first 2 years after a boreal wildfire. Most live tree residuals died during the first 2 years (63%), but the tree fall rate was much lower (15%). During this time, few snags (<1%) fell to become downed wood (data from Perera *et al.* 2011).

factors were shown by Dalziel and Perera (2009), who illustrated the importance of the scale of observation in describing the relationship between the survival of live tree residuals, fire intensity, and forest community structure, and concluded that simple observational studies may not be sufficient to explain the dynamics of live tree residuals after a fire.

Residual patches

There are many reports of the presence of residual patches within wildfire foot-prints, but these were based on different study durations: from a couple of years (e.g., Perera *et al.* 2009b), to a decade (e.g., Gluck and Rempel 1996, Andison 2012, Dragotescu and Kneeshaw 2012) or more than 75 years (e.g., Arseneault 2001). From these results, it appears that the broad-scale patterns of residual patches may not change much over time. However, it is possible that the area of each patch may begin to contract immediately after the fire. We based this hypothesis on our observations that the edges of residual patches are highly dynamic – there is much windthrow in the live trees at the exposed borders of residual patches, and there is also a high degree of delayed mortality in these transitional zones between burned and unburned areas (Figure 3.13).

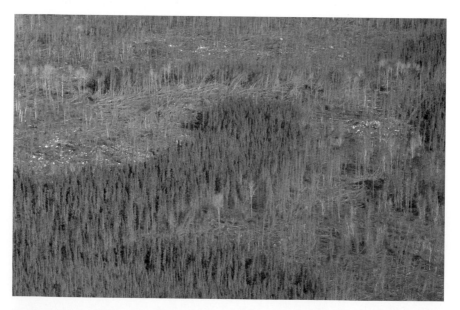

Figure 3.13   Evidence of windthrow at the edges of residual patches in a 2-year-old boreal wildfire footprint.

Toward improved definitions and assessment

As the examples and discussion in the previous section show, there is considerable variability in how researchers have considered, defined, and studied residuals. The variation makes it difficult to establish a coherent picture of how these residuals form and change over time. In this section, we propose possible improvements in approaches and methods, with the goal of providing a more holistic understanding of wildfire residuals.

Reasons for improvement

Based on an extensive review of the scientific literature, we found some general trends with respect to how researchers have perceived, described, and studied boreal wildfire residuals. These trends are not inconsequential because they affect how the structure of the residuals is portrayed, understood, and valued.

First, we found that researchers have generally considered residuals from a narrow perspective, with some specific utility in mind, rather than holistically. The definitions, studies, and estimates of the extent of residuals have all suffered from this mindset, which has led to the present incomplete understanding of residuals. Residuals have typically been studied with a reductionist approach, and assessments have often focused on one particular component or type of residual in isolation, providing information about only one aspect of the post-fire system at a given point in time. This narrow focus may serve to answer a specific research question, but it often means that descriptions of residuals are not provided within the context of the larger system. For example, studies may report partial information about residual abundance by focusing only on large stems, or may ignore components other than snags and live trees, such as the forest floor and understory. Study approaches that lack a holistic perspective of what fire leaves behind and how those residuals may change over time, hamper our understanding of post-fire systems and their recovery.

Secondly, we found wide variation in the definitions, scales, and methods that have been used to investigate residuals. Reference to the "dead" and "live" residual structure in the literature is often ambiguous. How the residuals formed (i.e., whether or not they were affected by the fire) is often not differentiated. In most cases, when residuals are referred to as "dead", the tree stems that had died before the fire are not distinguished from those that were killed by the fire. Similarly, so-called "live" residuals may only be alive at the time of the study, may soon disappear due to the generally high rates of delayed mortality, and may include stems that were and were not influenced by fire. Almost all studies have defined residuals somewhat differently, both qualitatively (e.g., what constitutes a residual patch) and quantitatively (e.g., the dimensions of residual snags), with little regard for the definitions used by other researchers.

The same trend can be seen in the scales and methods used in various investigations; for example, there is wide variation among studies in the spatial resolution of the surveys, the timing of assessment after fire, and the methods of assessing and reporting the characteristics of the wildfire footprint. These factors appear to vary from study to study, with no attempt made to standardize them, and the parameters thus employed are highly subjective rather than embedded in any broader concept of the ecology of residuals.

An additional common problem is the variation in the base used to estimate the proportion of residual patches within a fire. For example, the area of residual patches has been reported as a proportion of the total fire area (i.e., the area bounded by the external fire perimeter), of the total burnable area (i.e., the area of all cover types that did burn and that could have burned, excluding bodies of water and areas with no fuel, such as exposed bedrock), and of the total burned area (i.e., the area burned within the fire perimeter). This is sometimes a result of how fire perimeters were delineated, and other times results from a choice made by the researcher – but in either case, the different base for calculation introduces substantial differences in estimates of the proportions of residual patches. This variation results purely from the choice of estimation method, rather than from any inherent difference in the actual proportion.

In addition, there appears to be a disparity in terms of what information is reported after a given study; for example, there are many instances in the literature on residual ecology in which researchers have not reported important information, such as the number of fires studied, the time elapsed after the fire, where in the fires the surveys were conducted, and whether the fires were suppressed. In addition to the problems caused by differences in definitions and methods, researchers often fail to explicitly state their assumptions and describe their methods, leaving readers to guess how the estimates were derived. All these problems make it difficult to generalize and synthesize the published information.

Thirdly, we found a tendency to overgeneralize the results of studies of residual ecology, especially in terms of the spatial association of residuals with local environmental variables. Although some correlations may be evident in one fire or a small number of fires, the complexity of the factors that govern the origin of wildfire residuals (as we discussed in detail in Chapter 2) make it unlikely that such local trends represent global phenomena. For example, researchers commonly claim that residual patches occur mostly around bodies of water, and especially on their leeward side. Although such correlations represent suggestions or hypotheses for subsequent study, they are rarely treated as such in the literature. Instead these claims are perceived as general trends and are not rigorously tested as *a priori* hypotheses in new investigations. Furthermore, determining spatial associations between residuals and characteristics of both the site and the fire are not based on an explicit recognition that environmental variables are nested within weather patterns and other contextual variables responsible for generating and driving a given fire. Furthermore, many generalized associations extended or postulated have been based on study samples from a single or a few fires.

To overcome these and other limitations, and to progressively improve our knowledge of the ecology of residuals, we propose that studies of residuals should be standardized. Toward this goal, we describe a broad framework that begins with objective definitions and terms, is followed by suggestions for approaches for assessing wildfire residuals holistically within the overall context of the fire and its spatial footprint, and concludes with a list of the minimum information that should be obtained and reported in all studies of residuals.

Definitions of wildfire residuals

As in other fields of ecological study, a proliferation of terminology, its indiscriminate use, and the resulting ambiguity and confusion is evident in the scientific literature on boreal wildfire residuals. As discussed earlier, the range of terms used is extensive (Table 3.2). Although this variation may be unavoidable because of differing study objectives, practical necessities, and other circumstances, we suggest that adopting a more consistent conceptual premise founded on simplified principles of fire behavior and on the response of vegetation to fire will provide a robust foundation for standardizing the definitions and terminology associated with wildfire residuals. As a wildfire spreads through what will eventually become the fire footprint, the fire front avoids certain areas due to the effects of local fire weather, differences in vegetation and site conditions, or both. This dichotomy in fire behavior between burned and skipped areas provides a logical foundation for defining residuals: vegetation that escaped the fire and vegetation that did not. This basis is similar to the conceptual description of plant responses to fire provided by Rowe (1983).

When fire does reach a specific location, all vegetation within the fire footprint is affected to some degree: thus, our main categories of residuals are *unburned* and *burned*. This distinction is important because not only are the origins of these residual categories different, but their contributions to post-fire ecosystem recovery will also differ (as we discuss in Chapter 4). This perspective and definition will depend on the scale of the observations. When viewed at the level of an entire fire footprint (i.e., from the air), burned areas will appear as a collection of partially or completely scorched crowns. However, that same area viewed at the ground level will reveal that, in addition to the dead stems, both standing and downed, and in addition to the areas with a charred understory or a bare

Even though we focus mostly on post-fire tree residuals in this book, all biomass that is left behind within a wildfire footprint, without being reduced to ashes and gaseous emissions – whether dead or alive, plant or animal, and directly affected by fire or not – are residuals of the fire event in a broader sense. Such remnants of the pre-burn ecosystem are evident aboveground, on the forest floor, and belowground. The aboveground residual biomass includes taller vegetation – trees and shrubs – and animals ranging from invertebrates to birds and mammals.

The residual pool on the forest floor consists of woody biomass such as fallen tree stems, stumps, and branches; herbaceous biomass such as grasses, mosses, and lichens; fungal species; and litter and other plant debris. It also includes live vertebrates and invertebrates such as small mammals, reptiles, and arthropods.

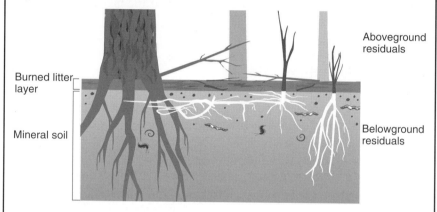

Aboveground
residuals

Burned litter
layer

Mineral soil

Belowground
residuals

The belowground residual biomass contains roots of all plant species, buried plant parts, fungal species, and live animal species such as subterranean arthropods, annelids, and nematodes. This reservoir of belowground residuals is substantial, but has not garnered much attention from researchers. Consequently, much less is known about its structure and function.

(a)

The species that form the soil seedbank in boreal forests include the shrubs pin cherry (*Prunus pensylvanica*), raspberry (*Rubus* spp.) (a), and currant (*Ribes* spp.); the herbaceous ground cover geranium (*Geranium* spp.) (b); and sedges (Carex spp.).

TYPES OF WILDFIRE RESIDUALS AND THEIR EXTENT

(*continued*)

Seed in the seed bank may remain viable for years to decades, depending on the species, and germinate in response to the increase in light, warmth, and available moisture after a fire that removes the surface litter and duff, exposing mineral soil.

(b)

(c)

The bank of vegetative propagules that exists below the surface is less well known. It includes the rhizomatous roots of some species, such as the shrubs alder (*Alnus* spp.) and blueberry (*Vaccinium* spp.) (c), the herb bunchberry (*Cornus canadensis*), and the ferns bracken fern (*Pteridium aquilinum*) (d) and horsetail (*Equisetum* spp.). These propagules survive by being buried at least 5 cm below the surface, where they are well insulated from the heat of fire.

(d)

Some species, such as the shrub red-osier dogwood (*Cornus stolonifera*), may regrow both from sprouts from surviving roots or stolons or from seed in the soil seedbank following fire.

These belowground residuals – both the vegetative propagules and the soil seedbank – contribute considerably to the recovery of burned ecosystems.

forest floor, there will be areas that escaped the fire at a smaller scale. Such areas will contain live understory and forest floor vegetation. Thus, unburned areas, which are intact remnants of the pre-fire system, can occur at all scales, ranging from contiguous patches of forest at the tree level to small patches of ground vegetation and moss on the forest floor at finer scales. Although unaffected directly by the fire, these areas have become spatially isolated because of the surrounding burned area. The scale of observation is therefore an important *a priori* consideration in how residuals are described and defined.

We categorize burned residuals further as having either died or survived their exposure to the fire. Residuals that died during the fire include standing dead stems (snags) and downed dead stems. These two forms of wildfire residuals differ not only in their structure but also in their function. For avian species, for example, standing dead stems provide structure for perchers and climbers, whereas downed stems provide shade and detritus for ground-level foragers. Also, the rates of decay and the contributions to nutrient cycles will differ between standing and downed stems. The structure of dead residuals is not limited to the aboveground biomass but includes the charred stems of understory vegetation as well as the subterranean root systems of all dead vegetation. The probability of occurrence of dead residuals is most likely in areas of high fire severity and least likely in areas of low fire severity (Table 2.5 in Chapter 2). A special case of wildfire residuals includes the stems, whether standing or downed, that had died before the fire, as these differ from the fire-killed stems. The differentiation is ecologically important: these stems have different moisture levels and will therefore have different rates of fall and decay, and will not contribute equally to regeneration of the burned areas.

Burned areas also contain residual vegetation that survived the fire. These survivors are there because, as Rowe (1983) noted, they are "tolerants;" that is, they have either endured or resisted the lethal effects of fire through their various adaptations. "Endurers" include species that resprout from buried meristems, such as trembling aspen, alder, and blueberries (*Vaccinium* spp.); "resistors" are species that have thick bark and high crowns with minimal ladder fuel when they are mature, such as jack pine and lodgepole pine (*Pinus contorta*). Depending on the scale of observation and the biota of interest, the survivors may include trees as the overstory component, shrubs as the understory component, or forest-floor components such as patches of moss. Survivors may occur in clusters or as individuals. The probability of occurrence of surviving residuals is most likely in areas of low fire severity and least likely in areas of high fire severity (Table 2.5 in Chapter 2). Such clusters of live residual vegetation in burned areas can be easily differentiated from unburned residuals because they show evidence of fire, such as scorched foliage or stems. In contrast, unburned areas show, by definition, no signs of fire.

Unburned residuals are typically found in contiguous areas, which vary in size, shape, extent, and spatial distribution within a fire footprint. Therefore, they may be seen as islands of remnants in a matrix of completely or partially burned vegetation. Such islands may include more than one vegetation type, depending on their extent and location. These unburned residual patches may vary from long strips of tree cover along the edge of a lake to small inclusions in low-lying areas or to wetlands and larger inclusions that are not associated with any particular landform

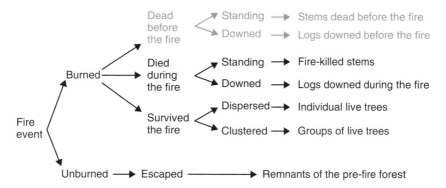

Figure 3.14   A schematic diagram illustrating the categories of tree-level fire residuals, differentiated by whether they were affected by fire and, if affected, whether they survived or died, and their post-fire state. Although stems that died before the fire (shown in grey) do contribute to the post-fire pool of dead wood, they should be distinguished from fire-created dead residuals.

feature. Unburned residuals may also be found as individuals, albeit infrequently. For example, a lone tree may escape the fire and survive among others that died. Although contiguous patches of unburned residuals will only occur in unburned areas of the fire footprint, individual unburned residuals (referred to as "escapees") may be found in areas of low to moderate fire severity.

These definitions of the residual categories are summarized from a tree-level perspective, based on their origin, in Figure 3.14. Similar keys could be developed for residual vegetation in the understory or on the forest floor.

We do not prescribe the use of any specific terms for each of these categories, as the optimal terminology will depend on the specific research goals and other considerations. However, we encourage researchers to nest the terminology that they adopt to describe wildfire residuals within the framework provided in Figure 3.14. For example, in this book we have focused primarily on tree-level residuals, and have used the following terms in the context of this framework:

- Stems that were dead before the fire, but that burned, are termed *pre-fire snags* and *pre-fire downed wood*, depending on whether they were standing or downed before the fire. These stems can be distinguished from those killed by the fire because they are generally burned more than adjacent stems due to their lower moisture content compared with stems that were alive before the fire. Pre-fire snags that burn and fall during the fire are thus included in the post-fire downed wood, as they fell during or after the fire.
- Stems that were alive before the fire and that died as a result of heat damage or burning are termed *snag residuals* and *fire-created downed wood*, again

depending on whether they are standing or downed, respectively, after the fire. The downed category includes live trees that fell during the fire.

- Stems that were subjected to the fire but survived and retain some green foliage immediately after the fire are termed *live tree residuals*. These stems can occur as individuals dispersed throughout the burned area or as clusters of living stems, but in both cases they show evidence of fire: the stems have some degree of charring and the crown shows some degree of scorching.

- Areas within the fire footprint that did not burn, and that show no evidence of fire damage but are surrounded by burned areas, are termed *residual patches*. These are intact remnants of the pre-fire system that include *trees*, dead standing stems (*snags*), and dead wood on the forest floor (*downed wood*).

These descriptions of tree-level wildfire residuals are summarized in Table 3.6, and illustrated in Figure 3.15, Figure 3.16, Figure 3.17 and Figure 3.18.

Table 3.6 Terms for describing tree-level wildfire residuals that will be used throughout the remainder of this book.

Origin	Term	Description
Died before the fire	Pre-fire snag	Standing stems that were dead before the fire and then burned
	Pre-fire downed wood	Logs that had fallen before the fire and then burned; includes pre-fire snags that burned and fell
Died during the fire	Snag residual	Trees that died as a result of being burned; stems have no green foliage
	Fire-created downed wood	Logs that burned and fell during the fire; includes live trees that died and fell
Survived the fire	Live tree residual	Standing stems that were subjected to the fire but that remained alive immediately after the fire (i.e., they retained green foliage); these can occur individually or in clusters
Escaped the fire	Residual patch	Unburned areas within the fire footprint that show no evidence of fire (i.e., they are remnants of the pre-fire forest that contain both live and dead stems, with the latter including standing and downed stems)

(a)

(b)

(c)

Figure 3.15   Residual stems that had died before the fire include (a and b) *pre-fire snags* and (c) *pre-fire downed wood*. These stems typically have little or no remaining bark and are charred more severely than the stems that were alive before the fire; roots of pre-fire downed wood are burned.

(a)

(b)

Figure 3.16   Residual stems that died during the fire include (a) *snag residuals* and (b) *fire-created downed wood*. These stems are typically less charred than the stems that were dead before the fire, and include live trees that fell during or shortly after the fire.

(a)   (b)

Figure 3.17 Trees that are alive soon after the fire, but that occur in burned areas, are termed *live tree residuals*. These may occur (a) individually or (b) in clusters. Because they occur in areas that did not escape fire, these stems may have signs of scorching on their bark or crown. Live tree clusters are distinct from residual patches because the latter show no evidence of fire.

In time, live tree residuals, pre-fire snags, and snag residuals will all add to the pool of downed wood. Residual patches may change in shape and total area due to the fall of trees and snags along the edges of the burned areas, but will remain evident for many years after the fire.

Improved study approaches

Instead of describing individual study methods in isolation, we discuss how wildfire residuals can be assessed by studying the fire footprint and fire severity first, and then studying the residual types nested within those fire severity patterns. This approach accommodates subjective assessments of residuals, but ensures that residuals are studied in the context of how they formed (and therefore how they will function ecologically), while recognizing the scale of the assessment. We believe that adhering to this conceptual framework will facilitate comparisons among studies and will therefore enable future syntheses and generalizations of trends.

Ideally, any study of wildfire residuals must begin by quantifying the character-istics of the fire's spatial footprint. This first step involves identifying the external perimeter of the fire, so that the extent of the fire and its constituent land cover

(a)

(b)

Figure 3.18 *Residual patches* are contiguous areas that have escaped the fire (lighter areas) and that show no evidence of fire, in contrast to areas that are burned (darker areas). These patches occur throughout most fire footprints and are intact remnants of the pre-burn ecosystem.

types can be estimated. The methods used to map fire footprints include manual tracing (airborne or ground level), GPS tracking (airborne or ground level), and interpreting remote-sensing data such as aerial photographs or satellite imagery. The choice of the method influences, among other aspects of the study, the spatial resolution by which the fire perimeter and its contents can be delineated. At present, there is no standardized approach to determining the fire perimeter; in fact, in most of the scientific literature no specific details about how the perimeter was determined are mentioned. As a solution, recent suggestions are to use clear and repeatable methods (Remmel and Perera 2009, Andison 2012). Regardless of the approach, the spatial resolution (i.e., the minimum unit of area that can be mapped confidently) will determine where the fire perimeter is located and which discontinuous satellite fires caused by spotting should be included within that perimeter (Remmel and Perera 2009). Different fire mapping choices can result in varied estimates of perimeters and extents for the same fire (Figure 3.19), and therefore different assessments of the characteristics of its residuals.

The next step is mapping the fire severity within the footprint. By *fire severity*, we mean the pattern of damage that is evident immediately after the fire, not the *burn severity*, which reflects long-term consequences that will eventually become apparent (French *et al.* 2008, Keeley 2009). As we discussed in Chapter 2, fire severity is clearly a continuum, but in practice it is treated as a discrete variable in most applications. Researchers can define broad classes of fire severity (Table 2.5 in Chapter 2) and map those classes using satellite imagery. An increasing number of types of satellite imagery, with higher spatial resolutions and more detailed spectral information, are becoming available for fire severity mapping in boreal forest landscapes, and they are accompanied by increasingly advanced analytical techniques. Therefore, the ability to map fire severity is likely to become more effective and widely acceptable (French *et al.* 2008).

Once the fire severity classes have been mapped, it is easy to estimate the locations and extent of unburned residual areas at the native resolution of the data. Only subsequently should researchers apply minimum-size filters and other restrictive definitions to characterize contiguous areas of unburned residuals (i.e., residual patches), and these constraints should be based on specific study questions. This approach preserves the information on unburned residual areas that fall below the minimum-size filters, and which would otherwise be lost if residual patches were mapped as the first step. From this starting point, other residual categories (Figure 3.14) can be defined and mapped within the respective fire severity classes, whether based on aerial photographs or ground surveys or a nested combination of the two (e.g., Perera *et al.* 2009a). This approach avoids the common problem of confounding clusters of live tree residuals, which occur in low- and moderate-severity classes, with unburned residual patches. In addition, if feasible, it is useful to define criteria for including residual stems in calculations

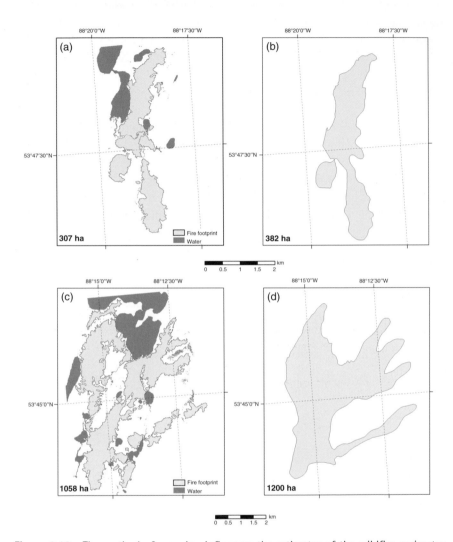

Figure 3.19 The method of mapping influences the estimates of the wildfire perimeter (the external boundary of the footprint) and the footprint (the spatial signature of the fire, which represents the area bounded by the perimeter). For example, wildfire footprints (a) and (c) are derived from IKONOS satellite imagery, analyzed at an 8-m resolution, whereas (b) and (d) were manually traced from 1:20 000 aerial photos mapped by Ontario Ministry of Natural Resources (Perera *et al.* 2009b, reproduced with permission).

of the abundance and extent of residuals. As with residual patches, minimum-size filters (i.e., diameter and height) should be applied only after this initial classification, so as to preserve information on the total number of residual stems.

Equally important is the timing of sampling. Because residuals, and especially live trees and patches, are highly dynamic and change rapidly during the first several years after a fire, timing is critical for estimating their extent and abundance. It is also preferable to include more than one fire in studies of residuals, unless the intent is simply exploratory. The among-fire variability in the extent of residuals could be extremely high under boreal conditions, for the numerous reasons discussed in Chapter 2. This variability must be captured by sampling more than one fire, especially if conclusions will be derived about the spatial patterns of the residual structure and its associations with fire and site characteristics.

In addition to adopting a robust approach to studying residuals (Figure 3.20), other considerations are needed to ensure that the study results can be readily interpreted, compared with other studies, and synthesized. For example, reporting the extent of residual patches as a proportion of the burnable area of the fire (i.e., excluding unburnable areas such as bodies of water and exposed bedrock from the total area within the fire perimeter) removes the inaccuracy that results from including these areas in the total. Similarly, reporting the abundance of live trees and snag residuals as a proportion of the total stem count provides for easier comparison than reporting only the absolute number or the density of stems. To this end, we propose a list of information (Table 3.7) that must be considered as the minimum requirements when reporting the extent and patterns of wildfire residuals.

Finally, the continuation of surveying residuals and developing *post hoc* reconstructions of scenarios must be reserved only for exploratory studies. Spatial pattern analyses of boreal wildfire residuals must evolve towards statistically robust testing of hypotheses that are conceptually based on the mechanisms of residual formation and informed by the results of previous studies, leading to progressive development and refinement of new hypotheses. This approach will advance the understanding of residuals and make it possible to compile a coherent body of reliable knowledge.

Summary

In this section, we summarize the range of views and descriptions of wildfire residuals in boreal forests and their extent, based on published information. Much of this body of knowledge has been derived from efforts to study wildfire residuals in relation to their ecological values and, to a lesser extent, from surveys of post-fire patterns in boreal forest landscapes.

TYPES OF WILDFIRE RESIDUALS AND THEIR EXTENT

TYPES OF WILDFIRE RESIDUALS AND THEIR EXTENT

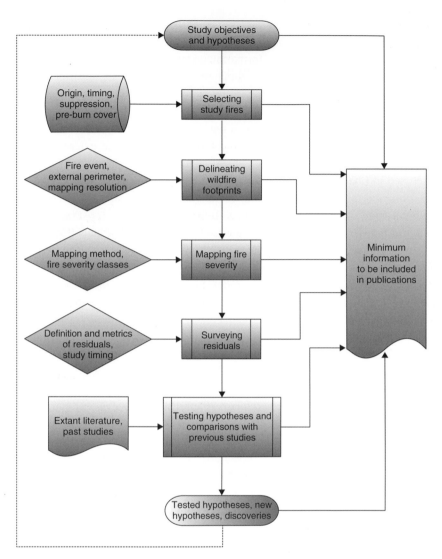

Figure 3.20   A stepwise framework proposed for studying wildfire residuals to ensure progressive reduction of uncertainties, and the advancement of scientific knowledge. The minimum information criteria are detailed in Table 3.7.

Three major types of residuals have been recognized in wildfire footprints: *residual patches* comprise spatially contiguous clusters of live trees, *live tree residuals* comprise spatially dispersed individual live trees, and *snag residuals* comprise dead trees that occur in clusters or individually. *Downed logs* are a sub-type of snag residual. As shown earlier (Table 3.2), the terminology used to describe these categories varies widely among studies and publications, as do the metrics used

Table 3.7 Recommendations for the minimum information requirements when reporting on wildfire residuals.

Aspect	Information needed	What to report
Fire	How many and where?	• Number and locations of fires
	When burned?	• Ignition source, timing (year, season), and duration of each fire
	How sampled?	• Specific approach used to map the fire perimeter, including the spatial resolution and minimum mapping unit • How the degree of fire severity is distinguished and defined (an objective, quantitative approach is preferable)
	What burned?	• Burnable area within the fire footprint, excluding water, bedrock, and other non-burnable area
Residuals	Why sampled?	• Specific goals of studying wildfire residuals • Specific hypotheses, formulated based on the formation mechanism, other scientific premises, and results of previous studies
	When sampled?	• Time elapsed since the fire (e.g., months or years) at the time of the study
	Where sampled?	• What parts of the fire were sampled? If only a selected part of fire, why and how was that part selected?
	How defined?	• Explicit definitions of residual categories and entities within a hierarchical framework defined based on the formation mechanism • Minimum-size filters and metrics used to estimate abundance and extent
	How studied?	• Method or methods used to assess (enumerate, locate) residual entities with sufficient detail to facilitate comparisons among studies
	What discovered?	• Estimates of residual abundance in the context of the pre-fire system • How findings are related to previous studies of residuals • Results of hypotheses that were tested, and findings that advance the empirical information and scientific knowledge of residual ecology

TYPES OF WILDFIRE RESIDUALS AND THEIR EXTENT

to define and quantify them (Figure 3.7). The effect of this inconsistency is not only semantic; it is likely to confuse first-time readers and students, and complicates among-study comparisons even for an expert. Therefore, we have proposed a nested framework for defining boreal wildfire residuals (Figure 3.14) that acknowledges the overall heterogeneity within a wildfire footprint, and that is founded on how residuals form in response to variations in fire severity and on how they will function ecologically, especially during the recovery of the post-fire ecosystem.

In essence, we suggest that residuals should be categorized as those that (a) escaped the fire (residual patches), (b) survived exposure to the fire (individual and clusters of live tree residuals), and (c) died as a result of exposure to the fire (snag residuals and downed logs). The stems that died *before the fire* are thus separated from those killed *during the fire*. This framework allows researchers to use their own definitions based on their study goals, while ensuring that their definitions can be readily related to the definitions used by others.

Boreal wildfire footprints contain residuals from all of these categories; that is, our literature review revealed no reports of the absence of any of these categories when an entire wildfire footprint was assessed. However, reports of the abundance or extent of residuals vary widely (Table 3.8). In part, this is due to the high heterogeneity both among and within boreal wildfires. In addition, it is extraordinarily difficult to describe general trends in the extents and patterns of residuals because the definitions of residual categories, the dimensions used to quantify the residuals, the sampling approaches, and the research methods all differ extensively among investigations and, in many instances, have not been explicitly reported. The available information on the abundance and extent of residuals also suffers from inadequate sample sizes because of the inaccessibility of fire locations (making replication difficult), the unpredictability of fire occurrence (leading to a need for opportunistic rather than pre-planned studies), and the absence of complete data on the pre-fire vegetation.

As well, differences in the timing of investigations (the delay after the fire) confound the results, because most such studies do not capture the highly dynamic

Table 3.8 Summary of extent/abundance of the major types of boreal wildfire residuals as reported in the literature.

Residual category	Abundance/extent
Snag residuals	238–2400 stems ha^{-1}
Live tree residuals	0–697 stems ha^{-1}
Residual patches	0–22% of burned area

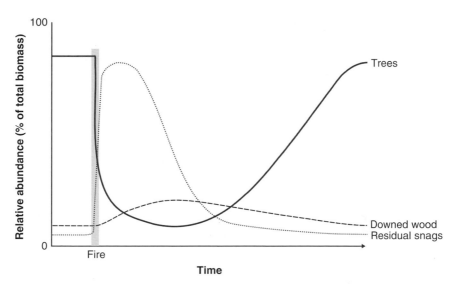

Figure 3.21   Generalized illustration of the changes in relative abundance of live tree and snag residual biomass in a burned area.

nature of the residual structure. This is especially so during the first few years after the fire, when there is a high degree of transition due to windthrow and delayed mortality among the live trees that initially survived the fire. As illustrated in Figure 3.21, live tree residuals decrease rapidly in the immediate post-fire period, increasing the abundance of snag residuals and downed wood. Snag residuals decline more gradually, contributing downed wood, which slowly decays. The exact timing of these pulses and the longer-term fluctuations are not known. They may be influenced by the fire severity, as determined by a combination of fire weather, vegetation composition, and local geo-environmental factors, and therefore vary both within and among fire events. Nonetheless, these ongoing dynamics in wild-fire residuals make timing a key factor in studies of residual abundance and extent. Given that they are highly heterogeneous spatially and dynamic temporally, we suggest that future studies of wildfire residuals adhere to a set of standards, such as those proposed in this chapter, to improve the quality and comparability of the results of their assessments.

As a starting point for the development of such standards, we recommend that researchers adopt the following structure for their investigations:

1. Clearly define the types of residuals based on their origins and formation mechanisms (e.g., using the categories defined in this chapter).
2. Develop robust and statistically valid hypotheses based on data from previous research before beginning to collect data.

3. Choose a consistent set of dimensions as the threshold for including something as a residual.
4. Obtain as much information as possible on the pre-fire and post-fire environmental and topographic conditions.
5. Choose study times that account for the post-fire dynamics of residuals (i.e., to capture the rapid changes that occur within the first few years after a fire as well as longer-term changes).

In general, investigations that consistently define residuals in relation to their formation, the dimensions that quantify residuals, and the choice of sampling and analytical methods based on previous studies; that assemble complete contextual information; and that judiciously select study times will make the knowledge of residuals more certain.

The presence and extent of residuals are commonly ascribed to variables such as the age and species of trees, fuel barriers and topography, and location. Most of these assertions have been based on correlations found within wildfire footprints that are based on single-scale and exploratory data analyses, and have relied on incomplete data about the factors that govern the formation of residuals. Although some researchers have made it clear that these are only observations and must be treated as hypotheses for future investigation, many of these spatial associations have become dogmas, such as claims that "residual patches are always found in … " or that " … tree species X produces the most live residuals". As we discuss in Chapter 5, these beliefs, however uncertain or unfounded, can dictate how boreal forests are managed at vast scales; therefore, errors and imprecision in these beliefs can have far-reaching consequences.

Researchers can minimize these uncertainties, and can attempt to determine the causes behind the spatial patterns of occurrence of residuals within wildfire footprints, only by adopting rigorous approaches. For example, previous findings and knowledge of residual formation should be used to formulate *a priori* hypotheses that will guide the investigation of potential explanatory factors. In addition, these factors should be examined at multiple scales. The resulting approach defines a nested hierarchy of factors, including all factors that govern the formation of residuals in analyses, and employs robust statistical techniques to rigorously analyze the collected data and test the hypotheses.

In sum, our knowledge of the types of residuals and their extent is in its infancy, and our understanding of this complex system is far from certain. The typical rigor and the meticulousness seen in ecological studies, such as those on ecological processes that we describe in Chapter 4, is not common in studies of residuals and their extents. This is partly because researchers have lacked robust and consistent definitions of the residuals created by boreal wildfires, and partly due to the wide range of motivations for studying residuals; most such assessments have treated residuals as only a peripheral aspect of the primary research. There is therefore much potential for improving the objectivity, consistency, and diligence in how residuals are considered and assessed, and hence much potential for reducing the uncertainties in this important area of ecological research. To work toward this goal, we hope that readers of this chapter will adopt and improve on our proposed classification for residuals and our framework for designing future research to study residuals.

References

Andison, D.W. (2012) The influence of wildfire boundary delineation on our understanding of burning patterns in the Alberta foothills. Canadian Journal of Forest Research 42, 1253–1263.

Angers, V.A., Gauthier, S., Drapeau, P., Jayen, K., and Bergeron, Y. (2011) Tree mortality and snag dynamics in North American boreal tree species after a wildfire: a long-term study. International Journal of Wildland Fire 20, 751–763.

Apfelbaum, S.I. and Haney, A. (1981) Bird populations before and after wildfire in a Great Lakes pine forest. Condor 83, 347–354.

Apfelbaum, S.I. and Haney, A. (1986) Changes in bird populations during succession following fire in the northern Great Lakes wilderness. In Lucas, R.C. (ed.) Proceedings – National Wilderness Research Conference: Current Research. USDA Forest Service, Intermountain Research Station, Ogden, UT. General Technical Report INT-212. pp. 10–16.

Arseneault, D. (2001) Impact of fire behavior on post-fire forest development in a homogeneous boreal landscape. Canadian Journal of Forest Research 31, 1367–1374.

Awada, T., Henebry, G.M., Redmann, R.E., and Sulistiyowati, H. (2004) Picea glauca dynamics and spatial pattern of seedling regeneration along a chronosequence in the mixedwood section of the boreal forest. Annals of Forest Science 61, 789–794.

Bergeron, Y., Leduc, A., Harvey, B.D., and Gauthier, S. 2002. Natural fire regime: a guide for sustainable management of the Canadian boreal forest. Silva Fennica 36, 81–95.

Beverley, J.L. and Martell, D.L. (2003) Modeling Pinus strobus mortality following prescribed fire in Quetico Provincial Park, northwestern Ontario. Canadian Journal of Forest Research 33, 740–751.

Bond-Lamberty, B.P., Wang, C., and Gower, S.T. (2003) Annual carbon flux from woody debris for a boreal black spruce fire chronosequence. Journal of Geophysical Research-Atmospheres 108(D23), 8220–8230.

Boulanger, Y., Sirois, L., and Hébert, C. (2011) Fire severity as a determinant factor of the decomposition rate of fire-killed black spruce in the northern boreal forest. Canadian Journal of Forest Research 41, 370–379.

Bradbury, S.M. (2006) Response of the post-fire bryophyte community to salvage logging in boreal mixedwood forests of northeastern Alberta, Canada. Forest Ecology and Management 234, 313–322.

Buech, R.R., Siderits, K., Radtke, R.E., Sheldon, H.L., and Elsing, D. (1977) Small mammal populations after a wildfire in northeast Minnesota. USDA Forest Service, North Central Forest Experiment Station, St. Paul, MN. Research Paper NC-151.

Carlson, D.J., Reich, P.B., and Frelich, L.E. (2011) Fine-scale heterogeneity in overstory composition contributes to heterogeneity of wildfire severity in southern boreal forest. Journal of Forest Research 16, 203–214.

Dalziel, B.D. and Perera, A.H. (2009) Tree mortality following boreal forest fires reveals scale-dependent interactions between community structure and fire intensity. Ecosystems 12, 973–981.

D'Amato, A.W., Fraver, S., Palik, B.J., Bradford, J.B., and Patty, L. (2011) Singular and interactive effects of blowdown, salvage logging, and wildfire in sub-boreal pine systems. Forest Ecology and Management 262, 2070–2078.

DeLong, S.C. and Kessler, W.B. (2000) Ecological characteristics of mature forest remnants left by wildfire. Forest Ecology and Management 131, 93–106.

Dragotescu, I. and Kneeshaw, D.D. (2012) A comparison of residual forest following fires and harvesting in boreal forests in Quebec, Canada. Silva Fennica 46(3), 365–376.

Eberhardt, K.E. and Woodard, P.M. (1987) Distribution of residual vegetation associated with large fires in Alberta. Canadian Journal of Forest Research 17, 1207–1212.

Epting, J. and Verbyla, D. (2005) Landscape-level interactions of prefire vegetation, burn severity, and postfire vegetation over a 16-year period in interior Alaska. Canadian Journal of Forest Research 35, 1367–1377.

Ferguson, S.H. and Elkie, P.C. 2003. Comparing snag abundance 20, 30, and 40 years following fire and harvesting disturbance in a boreal forest. Forestry Chronicle 79, 1–9.

Fisher, J.T. and Wilkinson, L. (2005) The response of mammals to forest fire and timber harvest in the North American boreal forest. Mammal Review 35, 51–81.

French, H.H.F., Kasischke, E.S., Hall, R.J. et al. (2008) Using Landsat data to assess fire and burn severity in the North American boreal forest region: an overview and summary of results. International Journal of Wildland Fire 17, 443–462.

Gandhi, K.J.K., Spence, J.R., Langor, D.W. et al. (2001) Fire residuals as habitat reserves for epigaeic beetles (Coleoptera: Carabidae and Staphylinidae). Biological Conservation 102, 131–141.

Gardiner, L.M. (1957) Deterioration of fire-killed pine in Ontario and the causal wood-boring beetles. The Canadian Entomologist 89, 241–263.

Gluck, M.J. and Rempel, R.S. (1996) Structural characteristics of post-wildfire and clearcut landscapes. Environmental Monitoring and Assessment 39, 435–450.

Graf, R., Bradley, M., Kearey, L., and Ellesworth, T. (2000) Successional responses of a black spruce ecosystem to wildfire in the Northwest Territories. Government of Northwest Territories, Department of Resources, Wildlife and Economic Development, Fort Smith, NWT. File Report No. 127.

Greene, D.F. and Johnson, E.A. (2000) Tree recruitment from burn edges. Canadian Journal of Forest Research 30(8), 1264–1274.

Haeussler, S. and Bergeron, Y. (2004) Range of variability in boreal aspen plant communities after wildfire and clear-cutting. Canadian Journal of Forest Research 34, 274–288.

Haines, D.A. (1982) Horizontal roll vortices and crown fires. Journal of Applied Meteorology 21, 751–763.

Harper, K.A., Bergeron, Y., Drapeau, P., Gauthier, S., and DeGrandpré, L. (2005) Structural development following fire in coniferous eastern boreal forest. Forest Ecology and Management 206, 293–306.

Harper, K.A., Lesieur, D., Bergeron, Y., and Drapeau, P. (2004) Forest structure and composition at young fire and cut edges in black spruce boreal forest. Canadian Journal of Forest Research 34, 289–302.

Hebblewhite, M., Munro, R.H., and Merrill, E.H. (2009) Trophic consequences of postfire logging in a wolf–ungulate system. Forest Ecology and Management 257, 1053-1062.

Hely, C., Bergeron, Y., and Flannigan, M.D. (2000) Coarse woody debris in the southeastern Canadian boreal forest: composition and load variations in relation to stand replacement. Canadian Journal of Forest Research, 30, 674–687.

Hely, C., Flannigan, M., and Bergeron, Y. (2003) Modeling tree mortality following wildfire in the southeastern Canadian mixed-wood boreal forest. Forest Science 49, 566–576.

Hobson, K. and Schieck, J. (1999) Changes in bird communities in boreal mixedwood forest: harvest and wildfire effects over 30 years. Ecological Applications 9, 849–863.

Hutto, R.L. (2006) Toward meaningful snag-management guidelines for postfire salvage logging in North American conifer forests. Conservation Biology 20, 984–993.

Imbeau, L., Savard, J.-P.L., and Gagnon, R. (1999) Comparing bird assemblages in successional black-spruce stands originating from fire and logging. Canadian Journal of Zoology 77, 1850–1860.

Kafka, V., Gauthier, S., and Bergeron, Y. (2001) Fire impacts and crowning in the boreal forest: study of a large wildfire in western Quebec. International Journal of Wildland Fire 10, 119–127.

Keeley, J.E. (2009) Fire intensity, fire severity and burn severity: a brief review and suggested usage. International Journal of Wildland Fire 18, 116–126.

Larsen, J.A. (1962) Major vegetation types of western Ontario and Manitoba from aerial photographs. University of Wisconsin, Department of Meteorology, Madison, WI. Technical Report No. 7.

Le Goff, H. and Sirois, L. (2004) Black spruce and jack pine dynamics simulated under varying fire cycles in northern boreal forest of Quebec, Canada. Canadian Journal of Forest Research 34, 2399–2409.

Lutz, H.J. (1956) Ecological effects of forest fires in the interior of Alaska. USDA Forest Service, Alaska Forest Research Center, Juneau, AK. Technical Bulletin 1133.

Madoui, A., Leduc, A., Gauthier, S., and Bergeron, Y. (2010) Spatial pattern analyses of post-fire residual stands in the black spruce boreal forest of western Quebec. International Journal of Wildland Fire 19, 1110–1126.

Manies, K.L., Harden, J.W., Bond-Lamberty, B.P., and O'Neill, K.P. (2005) Woody debris along an upland chronosequence in boreal Manitoba and its impact on long-term carbon storage. Canadian Journal of Forest Research 35, 472–482.

McIver, J.D. and Starr, L. (2001) A literature review on the environmental effects of postfire logging. Western Journal of Applied Forestry 16, 159–168.

Methven, I.R., van Wagner, C.E., and Stocks, B.J. (1975) The vegetation of four burned areas in northwestern Ontario. Canadian Forestry Service, Petawawa Forest Experiment Station, Chalk River, ON. Information Report PS-X-60.

Miyanishi, K. and Johnson, E.A. (2002) Process and patterns of duff consumption in the mixedwood boreal forest. Canadian Journal of Forest Research 32, 1285–1295.

Morissette, J.L., Cobb, T.P., Brigham, R.M., and James, P.C. (2002) The response of boreal forest songbird communities to fire and post-fire harvesting. Canadian Journal of Forest Research 32, 2169–2183.

Murphy, E.C. and Lehnhausen, W.A. (1998) Density and foraging ecology of woodpeckers following a stand-replacement fire. Journal of Wildlife Management 62, 1359–1372.

Nalder, I.A. and Wein, R.W. (1999) Long-term forest floor carbon dynamics after fire in upland boreal forests of western Canada. Global Biogeochemical Cycles 13, 951–969.

Nappi, A. and Drapeau, P. (2011) Pre-fire forest conditions and fire severity as determinants of the quality of burned forests for deadwood-dependent species: the case of the black-backed woodpecker. Canadian Journal of Forest Research 41, 994–1003.

Nappi, A., Drapeau, P., Giroux, J.-F. *et al.* (2003) Snag use by foraging black-backed woodpeckers (*Picoides arcticus*) in a recently burned eastern boreal forest. Auk 120, 505–511.

Nappi, A., Drapeau, P., Saint-Germain, M., and Angers, V.A. (2010) Effect of fire severity on long-term occupancy of burned boreal conifer forests by saproxylic insects and wood-foraging birds. International Journal of Wildland Fire 19, 500–511.

Nordin, J.O. and Grigal, D.F. (1976) Vegetation, site, and fire relationships within the area of the Little Sioux fire, northeastern Minnesota. Canadian Journal of Forest Research 6, 78–85.

Nowak, S., Kershaw, G.P. and Kershaw, L. (2002) Plant diversity and cover after wildfire on anthropogenically disturbed and undisturbed sites in subarctic upland *Picea mariana* forest. Arctic 55, 269–280.

Ohmann, L.F. and Grigal, D.F. (1979) Early revegetation and nutrient dynamics following the 1971 Little Sioux forest fire in northeastern Minnesota. Forest Science 25(4 Suppl.).

Perera, A.H., Buse, L.J., Routledge, R.G., Dalziel, B.D., and Smith, T. (2008). An assessment of tree, snag, and downed wood residuals in boreal fires in relation to Ontario's policy directions for emulating natural forest disturbance. Ontario Ministry of Natural Resources, Ontario Forest Research Institute, Sault Ste. Marie, ON. Forest Research Report No. 168.

Perera, A.H., Dalziel, B.D., Buse, L.J., and Routledge, R.G. (2009a) Spatial variability of stand-scale residuals in Ontario's boreal forest fires. Canadian Journal of Forest Research 39, 945–961.

Perera, A.H., Dalziel, B.D., Buse, L.J., Routledge, R.G., and Brienesse, M. (2011) What happens to tree residuals in boreal forest fires and what causes the changes? Ontario Ministry of Natural Resources, Ontario Forest Research Institute, Sault Ste. Marie, ON. Forest Research Report No. 174.

Perera, A.H., Remmel, T.K., Buse, L.J., and Ouellette, M.R. (2009b) An assessment of residual patches in boreal fire in relation to Ontario's policy directions for emulating natural forest disturbance. Ontario Ministry of Natural Resources, Ontario Forest Research Institute, Sault Ste. Marie, ON. Forest Research Report No. 169.

Pohl, G., Langor, D., Klimaszewski, J., Work, T., and Paquin, P. (2008) Rove beetles (Coleoptera: Staphylinidae) in northern Nearctic forests. The Canadian Entomologist 140, 415–436.

Rees, D.C. and Juday, G.P. (2002) Plant species diversity on logged versus burned sites in central Alaska. Forest Ecology and Management 155, 291–302.

Remmel, T.K. and Perera, A.H. (2009) Mapping natural phenomena: boreal forest fires with non-discrete boundaries. Cartographica 44, 275–289.

Richmond, H.A. and Lejeune, R.R. (1945) The deterioration of fire-killed white spruce by wood-boring insects in Northern Saskatchewan. Forestry Chronicle 21, 168–192.

Ross, D.A. (1960) Damage by long-horned wood borers in fire-killed white spruce, central British Columbia. Forestry Chronicle 36, 355–361.

Rowe, J.S. (1983) Concepts of fire effects on plant individuals and species. In Wein, R.W. and MacLean, D.A. (eds) The Role of Fire in Northern Circumpolar Ecosystems. Scope 18. John Wiley & Sons, Ltd, New York, NY. pp. 135–154.

Saint-Germain, M., Drapeau, P., and Hebert, C. (2004) Comparison of Coleoptera assemblages from a recently burned and unburned black spruce forests of northeastern North America. Biological Conservation 118, 583–592.

Saint-Germain, M. and Greene, D.F. (2009) Salvage logging in the boreal and cordilleran forests of Canada: integrating industrial and ecological concerns in management plans. Forestry Chronicle 85, 120–134.

Schaeffer, J.A. and Pruitt, W.O. (1991) Fire and woodland caribou in southeastern Manitoba. Wildlife Monographs 116, 1–39.

Schieck, J. and Hobson, K.A. (2000) Bird communities associated with live residual tree patches within cut blocks and burned habitat in mixedwood boreal forests. Canadian Journal of Forest Research 30, 1281–1295.

Schmiegelow, F.K.A., Stepnisky, D.P., Stambaugh, C.A., and Koivula, M. (2006) Reconciling salvage logging of boreal forests with a natural-disturbance management model. Conservation Biology 20, 971–983.

Schulte, L.A. and Niemi, G.J. (1998) Bird communities of early successional burned and logged forest. Journal of Wildlife Management 62, 1418–1429.

Schwab, F.E., Simon, N.P.P., Stryde, S.W., and Forbes, G.J. (2006) Effects of post-fire snag removal on breeding birds of western Labrador. Journal of Wildlife Management 70(5), 1464–1469.

Scotter, G.W. (1972) Fire as an ecological factor in boreal forest ecosystems of Canada. In Fire in the Environment: Symposium Proceedings. USDA Forest Service, Intermountain Forest and Range Experiment Station, Ogden, UT. Report No. FS-276. pp. 15–25.

Simard, A.J., Haines, D.A., Blank, R.W., and Frost, J.S. (1983) The Mack Lake Fire. USDA Forest Service, North Central Forest Experiment Station, St. Paul, MN. General Technical Report NC-83.

Simon, N.P.P., Stratton, C.B., Forbes, G.J., and Schwab, F.E. (2002) Similarity of small mammal abundance in post-fire and clearcut forests. Forest Ecology and Management 165, 163–172.

Slaughter, K.W., Grigal, D.F., and Ohmann, L.F. (1998) Carbon storage in southern boreal forests following fire. Scandinavian Journal of Forest Research 13, 119–127.

Stuart-Smith, K., Adams, I.T., and Larsen, K.W. (2002) Songbird communities in a pyrogenic habitat mosaic. International Journal of Wildland Fire 11, 75–84.

van Wagner, C.E. (1973) Height of crown scorch in forest fires. Canadian Journal of Forest Research 3, 373–378.

TYPES OF WILDFIRE RESIDUALS AND THEIR EXTENT

Viereck, L.A. and Dyrness, C.T. (1979) Ecological effects of the Wickersham Dome Fire near Fairbanks, Alaska. USDA Forest Service, Pacific Northwest Forest and Range Experiment Station, Portland, OR. General Technical Report PNW-90.

Wikars, L.-O. (2001) The wood-decaying fungus *Daldinia loculata* (Xylariaceae) as an indicator of fire-dependent insects. Ecological Bulletins 49, 263–268.

4 Ecological roles of wildfire residuals

Wildfires cause a drastic and sudden change in forest ecosystems, and many new ecological processes occur in the changed systems during their recovery from the disturbance. As discussed in the previous chapters, the changes caused by wildfires are not spatially uniform: post-fire ecosystems contain areas that are burned (to high, medium, and low severity) and unburned, all spatially juxtaposed. Consequently, not only do different ecological processes occur in these differently affected components of the system, but they may also interact spatially.

This chapter focuses on the diverse ecological processes, both biotic and abiotic, that occur within and among wildfire residuals at various spatial and temporal scales. We do not address all processes that occur in post-fire ecosystems, as that is beyond the scope of our focus on the role of residuals. For the reader interested in broader aspects of fire ecology, we recommend the texts by Chandler *et al.* (1983), Whelan (1995), and DeBano *et al.* (1998), and for boreal wildfires in particular, the text by Johnson (1992).

Ecology of Wildfire Residuals in Boreal Forests, First Edition. Ajith H. Perera and Lisa J. Buse.
© 2014 Ajith H. Perera and Lisa J. Buse. Published 2014 by John Wiley & Sons, Ltd.
Companion Website: www.wiley.com/go/perera/wildfire

As elaborated previously, our perception of residuals is broad: it includes all living vegetation remnants from the pre-fire ecosystem (e.g., trees, shrubs, mosses, lichens, fungi) as well as the vast pool of non-pyrolized aerial biomass (standing dead wood and downed wood). Our main focus in this chapter is trees, but we do consider other vegetation where possible. We discuss the role of these residuals from two perspectives: their *in situ* functions as remnants of the pre-fire ecosystem in which they continue to perform various ecological processes locally, such as acting as sources of regeneration, and their *ex situ* functions in which unburned vegetation plays important roles in adjacent burned areas by virtue of their proximity. Among the *ex situ* biotic ecological roles performed by unburned residuals, the most important is habitat provision for fauna that preferentially use these resources, that incidentally use residual patches as a temporary refuge, and that use the whole matrix of burned and unburned areas.

The roles of unburned residuals are particularly relevant during the early post-fire period, when they are critical in the recovery of the wildfire footprint by serving as sources of propagules that will help to renew the burned ecosystem. Unburned residuals at tree level include two components: areas that escape fire (residual patches) and stems that survive exposure to fire (live tree residuals). We do not address the ecology of live tree residuals specifically; this is because, given their transient nature and relative rarity of occurrence, live tree residuals have escaped attention of researchers and consequently not much is known about their ecological roles. In addition to the contributions to the wildfire footprint, we also consider the contribution of wildfire residuals to ecological processes that extend beyond the fire event, including those processes that occur at broader spatial scales in boreal forest landscapes and that influence regional biodiversity. Furthermore, we examine the roles of residual vegetation, both live and dead, in the carbon cycle and in sequestering terrestrial carbon and briefly discuss nutrient cycling and hydrological processes.

We first explore the biotic processes that occur in burned and unburned areas, snag residuals and residual patches, respectively, followed by the abiotic processes. In doing so, we rely on knowledge from published studies in boreal forest landscapes, most from North America and Fennoscandia, as information on Eurasian boreal forests is not readily available. As mentioned in Chapter 1, when information was lacking on specific topics, we occasionally use information from temperate forest ecosystems.

Our discussion begins with events that occur soon after a fire and ends when residuals no longer play a critical role in ecosystems that are recovering from a fire. This does not imply that the residuals cease to be ecologically important; on the contrary, the legacy effect they create could last for centuries, if not longer.

ECOLOGICAL ROLES OF WILDFIRE RESIDUALS

Ecological processes involving snag residuals

Wildfires occur over brief periods (usually hours or a few days), and destroy trees, understory vegetation, the forest floor, and most organisms within the affected area. In contrast, most other forest disturbances that cause tree mortality and create dead wood over large areas, such as insect epidemics, occur slowly, over weeks and months. In addition, the latter often affect only a single tree species or species group. Furthermore, in wildfires much of the forest ecosystem is transformed into dead wood in the form of standing or fallen stems. The most obvious feature of the immediate post-fire system is therefore the abundance of snag residuals (standing dead wood) as well as downed wood (coarse woody debris). At a glance, this ecosystem of standing and fallen dead stems may appear to be devoid of life (e.g., Figure 3.2 in Chapter 3), except for scattered live tree residuals that, as we discussed in Chapter 3, are transient, since they often quickly become snags or downed wood, and so are included with snag residuals in our discussion of ecological roles.

The burned forest and the soil surface do not stay sterile and lifeless for long. An interesting series of interrelated ecological events rapidly ensues, a process that some researchers consider unique to fire-driven forest landscapes such as the boreal forests of North America and Russia. The sequence starts with a mass-scale influx of insects into the burned forest, which triggers a parallel invasion of insectivorous birds. Some view this process and the ensuing patterns of insect and bird invasion to be an important component of maintaining biodiversity at regional scales. At the same time, the forest vegetation begins to re-establish from propagules that originate both within and outside the wildfire footprint.

Invasion by beetles

Most of the arthropods that inhabited either living trees (*arboreal* or tree-dwelling organisms) or the soil surface (*epigeic* organisms) in the pre-burn forest do not survive the heat of the fire (Saint-Germain *et al.* 2005, Pohl *et al.* 2008). However, this absence of life is brief. Within a couple of days after a sudden and intense wildfire disturbance, ecological processes in the burned system resume, and this resurgence is initiated by invasion of the dead and dying trees by insects, primarily beetles. These beetles can be divided into two groups based on their feeding preference (Saint-Germain and Greene 2009). Bark beetles primarily feed on the phloem and cambium that lie under the bark (e.g., *Acmaeops* spp., Coleoptera: Cerambycidae), and are therefore referred to as *phloeophagous*. Wood borers begin by feeding under the bark and subsequently bore into the wood to feed on the xylem (e.g., *Monochamus* spp., Coleoptera: Cerambycidae), and are therefore referred to as *xylophagous*.

Since both groups live in the woody (xylic) materials of dying or dead trees, they are called *saproxylic* species. Saproxylic beetles feed under the bark, drill into the wood, or feed on fungi (hyphae or sporocarps) that grow on dead or decaying wood. They depend on standing or fallen dead trees for at least part of their life cycle and, as noted by Grove (2002), they are a broad group and considered to be globally important.

Saproxylic beetle species are also among the most conspicuous and well-studied insects in post-fire boreal forests, both in Europe (e.g., Wikars 2002) and in North America (e.g., Boulanger and Sirois 2007, Saint-Germain *et al.* 2007). This group includes bark beetles such as *Ips* spp. (Coleoptera: Curculionidae) that prefer trees during the early stages of dying, and wood borers that prefer severely stressed trees that are nearly dead or recently killed. Both groups are sometimes called "stressed-host" species (Saint-Germain *et al.* 2007) because they either rarely infest vigorous trees or cause little damage when they do. Wood borers have been the most widely studied, especially sawyer beetles such as *Monochamus* spp. and metallic wood-boring beetles such as *Melanophila* spp. (Coleoptera: Buprestidae). Although researchers have found many specific exceptions and considerable variation in studies of saproxylic insects, here we follow the generalized sequence of post-fire beetle dynamics proposed by Boulanger and Sirois (2007), who note two distinct phases of post-fire colonization of residuals by saproxylic beetles in the Canadian boreal forest based on their studies in Québec (Canada).

First phase of the beetle invasion

The first phase consists of the beetle species that arrive immediately after fires to occupy recently killed, dying, weakened, or stressed trees. These species are termed *pyrophilous*, which means "fire-loving". Various signals attract the earliest pyrophilous species to post-fire sites. Vast smoke plumes from boreal wildfires (Figure 4.1) that disperse over long distances appear to be an important attractant for pyrophilous insects (Saint-Germain *et al.* 2004a); in fact, some pyrophilous beetles such as the black fire beetle (*Melanophila acuminata*) can detect smoke plumes from as far as 50 km (Evans 1966, Schütz *et al.* 1999). Such species, including the genera *Melanophila* and *Oxypteris* (Coleoptera: Buprestidae; Saint-Germain and Greene 2009), also appear to be attracted by the infrared radiation from wildfires (Evans 1964) and by volatile compounds that emanate from severely stressed or dying trees (Lindgren and Miller 2002). Not much is known with certainty about the mechanics of the invasion process, but some researchers (e.g., Saint-Germain *et al.* 2004c) hypothesize that the distance from an unburned forest, exposure to wind, and topography all influence how pyrophilous insect species colonize specific sites after a wildfire.

Wildfires are a frequent and widespread phenomenon in the boreal forest biome. Most wildfires are intense, especially those in North America. They spread across large areas, sending flames high into the forest canopy and consuming vast amounts of forest biomass. Ecological, social, and economic consequences of boreal wildfires are the focus of both scientific study and political debate. However, amid the general destruction that reduces a living forest to gases and ash, some elements of the forest structure remain intact.

Ecology of Wildfire Residuals in Boreal Forests, First Edition. Ajith H. Perera and Lisa J. Buse.
© 2014 Ajith H. Perera and Lisa J. Buse. Published 2014 by John Wiley & Sons, Ltd.
Companion Website: www.wiley.com/go/perera/wildfire

The vast pool of charred trees may look sterile, but they are important for many local and broader ecosystem processes. For example, immediately after the fire, a great wave of beetles arrives to feed and lay their eggs under the surface of the charred wood. A mass influx of woodpeckers then follows to feed on the beetle larvae. When the woodpeckers breed and depart for a new fire area, the cavities they vacate are occupied by other birds. During all of this activity, seed rains down from the cones of both burned and unburned trees, and new sprouts emerge from unburned roots and stems to help establish a new generation of vegetation.

Due to variations in the behavior of fire, many patches of vegetation escape fire damage and survive intact. These unburned patches range in scale from large areas of forest to clusters of understory vegetation to clumps of moss on the forest floor. Not only do they serve as a refuge for animals during the fire, but these patches also provide a source of biotic material for the recovering ecosystem in the short term and enhance biodiversity at genetic, species, and ecosystem levels to promote evolution in the long term.

While much remains to be explored further scientifically, it is clear that both burned and unburned residuals within wildfires contribute greatly to ecological processes across a range of spatial and temporal scales. The heterogeneity of patterns created in the landscape result in a montage of ecosystems of varying species and ages – a shifting spatial mosaic that is integral to, and defines, the boreal forest biome.

Figure 4.1   A large smoke plume rises from a boreal wildfire in northern Ontario (Canada). These smoke columns serve as an attractant for pyrophilous insects over long distances, sometimes as far as 50 km. (Reproduced with permission of Ontario Ministry of Natural Resources.)

ECOLOGICAL ROLES OF WILDFIRE RESIDUALS

During the first wave of the invasion, the burned area is usually filled with the noise of beetle foraging – the scraping and boring of wood – and beetle swarms can be seen everywhere. Their populations can be very large, for example, more than 300 larvae per square metre of snag surface (Saint-Germain *et al.* 2004b). As a result, the bases of snag residuals are covered with piles of wood residue produced by the beetles. These beetles are known to inhabit all coniferous species in the genera *Abies, Picea,* and *Pinus* (Saint-Germain and Greene 2009, Cobb *et al.* 2010), and we have observed vigorous wood borer activity in *Populus* spp. as well (Figure 4.2). The nature of the wood residue varies with the type of beetle: bark beetles do not produce conspicuous amounts of wood residue, but wood borers do. *Monochamus* spp., for example, produce quantities of stringy wood-chip residue that collects around tree bases, whereas buprestids produce fine wood dust, mostly compacted in the galleries under the bark. Added to these visible heaps of residue is insect frass, consisting primarily of larval excreta.

After their arrival in the fire footprint, adults mate and locate suitable host snag residuals in which to lay their eggs. Eggs are deposited within the bark of the snag residuals, and larvae typically hatch within 2 weeks (Rose 1957). As is typical for wood borers, both buprestid and cerambycid larvae feed on the phloem at first, and then move on to the xylem (Richmond and Lejeune 1945) by boring deep

(a)

(b)

(c)

(d)

Figure 4.2 Snag residuals of (a) jack pine (*Pinus banksiana*), (b) black spruce (*Picea mariana*), (c) balsam fir (*Abies balsamea*), and (d) poplar (*Populus* spp.) have their base surrounded by wood chips produced by the larvae of wood borers 2 months after a boreal wildfire. This is a typical scene in boreal forests during the first 2 years after a wildfire.

into the wood. This is where they overwinter and where they continue to feed and grow for a period of up to 2 years, moving between the wood and the surface to feed on phloem. Pupation occurs in enlarged galleries near the surface during the spring of the second year. Adults typically emerge in the summer – earlier at southern latitudes and later at northern latitudes (Gardiner 1957). These beetles do not persist for more than one generation (<2 years) on a specific host (Saint-Germain *et al.* 2007). Because the adults emerge in the summer and are short-lived, spring and summer fires that occurred before their emergence period provide a ready supply of stressed and dying trees. Snag residuals in fires that occur late in the fall are less susceptible to infestation by wood borers because adults that emerged during the summer would have already laid their eggs in trees killed by spring and summer fires. Consequently, fires that occur late in the fall produce trees that are likely to remain uninfested over the first winter and that may be attacked the following spring by bark beetles and then by shallow-boring cerambycids such as *Acmaeops* spp. (Gardiner 1957).

Typically, lower fire severity (for definition/description see Chapter 2) results in greater beetle infestation because uncharred trees house more beetles than completely charred trees (Richmond and Lejeune 1945). High beetle abundance in dying and highly stressed trees, especially in areas of lower fire severity, is likely due to higher stem moisture levels (Saint-Germain *et al.* 2004b), but it also could be that the nutritious cambium and phloem of these trees has not been destroyed. It appears that both bark and wood-boring beetles attack stressed stems of all sizes (Figure 4.3), but are more prevalent in larger stems, a pattern likely due to the larger surface area and the increased bark and phloem thickness, which provide more nutrients and moisture for these insects (Gardiner 1957, Saint-Germain *et al.* 2004c). Overall, it is clear that the beetles that colonize snag residuals have different feeding habits, and different preferences for the physical, chemical, and biological characteristics of the wood they inhabit, and thus differ in the timing of their occupancy of wood after a fire (Langor *et al.* 2006).

Although *Monochamus* and *Melanophila* are the most widely studied genera, other groups of beetles also invade shortly after a fire. The specific locations they colonize vary among species, and depend on geographic location and the time after fire when they are observed (Table 4.1). For example, Gardiner (1957) reported that as many as 24 beetle species – buprestids, pythids, cerambycids, curculionids, and scolytids – were present within the first few months after a boreal wildfire in Canada. Furthermore, the wood borers and bark beetles attract predatory beetles (e.g., Coleoptera: Carabidae), which are also early post-fire colonizers (Lindgren and Miller 2002). Predators such as rove beetles (Coleoptera: Staphylinidae) and checkered beetles (Coleoptera: Cleridae) arrive to feed on the saproxylic beetles (Wikars 2001, Boulanger and Sirois 2007, Toivanen and Kotiaho 2007). Because of this sudden influx of beetles, insect diversity increases rapidly in the burned areas

Two wood-boring beetles commonly found in snag residuals in boreal forests soon after the fire: (a) the black fire beetle (*Melanophila acuminata*) and (b) the white-spotted sawyer beetle (*Monochamus scutellatus*). The photos show (a, b) their adults, (c, d) their larvae, and (e, f) their feeding galleries, respectively. Black fire beetle larvae are smaller, with flat heads, and the sawyer beetle larvae are much larger, with round heads. The latter produce larger galleries in stems and more pronounced deposits of wood residue around tree bases.

Note that the images are not life-size or consistently scaled.

Reproduced with permission from (a) S. Luk; (b) M. Francis, Ontario Ministry of Natural Resources.

(a)   (b)

Figure 4.3   Wood-boring beetle activity is evident in snag residuals across a range of size classes in boreal wildfire footprints. The pocket knife in the foreground of both images is 10 cm long.

within days after the fire (Paquin 2008). Where fire severity is low, tree mortality occurs gradually, sometimes over several years, and this provides an extended supply of newly dead trees (e.g., Perera *et al.* 2011). This continuous recruitment of dead trees extends the presence of beetle species such as *Monochamus* (Nappi *et al.* 2010).

Second phase of the beetle invasion

Boulanger and Sirois (2007) mention a second "wave" of saproxylic insects that consists of *fungivores* (which consume fungi) and *saprophages* (which consume decaying wood), which invade as the snag residuals begin to fall. These beetles colonize the dead wood as the bark falls off and the subcortical feeders disappear. Insects in the second wave are called "dead-host" species, in contrast to the "stressed-host" species of the first wave of invaders that feed only on freshly killed trees (Saint-Germain *et al.* 2008). Several successive generations of dead-host species usually develop within the same host (Hanks 1999).

After the initial changes in beetle species composition, which occur within the first 2 years after the fire, changes over the next decade involve a decrease in the abundance of individuals of the subcortical saproxylic species. This decline

ECOLOGICAL ROLES OF WILDFIRE RESIDUALS

Table 4.1   Saproxylic beetles that invade post-fire snag residuals in boreal Fennoscandia and North America. Species names are provided for species known to be fire specialists or pyrophilous (*sensu* Evans 1964, 1966, Wikars 1997, 2001, Muona and Rutanen 1994). (Nomenclature is that used in cited papers.)

Family (common name)	Genus	Food category[a]	Timing (years after the fire)[b]		Source[c]	Location[d]
			0–1	2–5		
Bostrichidae (branch and twig borers)	*Stephanopachys substriatus*	NR	Y	Y	6, 7	F, S
Buprestidae (metallic wood-boring beetles, jewel beetles, or flat-headed borers)	*Anthaxia*	X	Y	Y	4	C
			N	Y	1	
	Chrysobothris	X	Y	N	1	C
		Ph			5	
	Melanophila acuminata	X	Y	NS	2, 7	C, S
	Melanophila	Ph	Y	Y	4	C
			Y	N	5	
	Oxypteris	X	Y	Y	1, 4	C
Cephaloidae (false longhorn beetles)	*Stenotrachelus aeneus*	X	Y	NS	7, 8	S
Cerambycidae (longhorned beetles or round-headed borers)	*Acmaeops proteus*	X	Y	Y	5	C
			Y	N	1	
	Acmaeops septentrionis	Ph	Y	NS	7, 8	S
	Anastrangalia	NR	Y	Y	3	C
	Arhopalus	X	Y	N	1	C
	Gnathacmaeops	X	Y	Y	1	C
	Monochamus scutellatus	Ph, X	Y	N	1	C
			Y	Y	5	C
	Neacanthocinus	Ph, X	Y	Y	5	C
	Pachyta	X	Y	Y	1	C
	Pogonocherus	Ph	Y	N	5	C
	Rhagium	X	Y	Y	1	C
	Tetropium	Ph, X	Y	Y	5	C
	Trachysida	NR	Y	N	1	C
Ciidae (minute tree fungus beetles)	*Cis* cf. *striolatus*	M	Y	Y	1	C
Corylophidae (minute fungus beetles)	*Orthoperus*	NR	Y	NS	6	F
	Sacium	M	Y	Y	1	C
Cryptophagidae (silken fungus beetles)	*Anchicera*	M	Y	Y	1	C
	Antherophagus	M	N	Y	1	C
	Atomaria	M	Y	Y	1, 3	C, F

(continued overleaf)

Table 4.1 (*continued*)

Family (common name)	Genus	Food category[a]	Timing (years after the fire)[b]		Source[c]	Location[d]
			0–1	2–5		
	Caenoscelis	M	Y	Y	1, 3	C, F
	Cryptophagus	M	Y	Y	3	F
	corticinus		Y	NS	7, 8	S
	Henoticus serratus	M	Y	Y	1, 3	C, F
			Y	NS	7, 8	S
Cucujidae (flat bark beetles)	*Laemophloeus*	P, M	Y	NS	7, 8	S
	muticus		Y	Y	3	F
	Pediacus fuscus	P	Y	Y	1, 3	C, F
			Y	NS	7	S
Curculionidae (snout and bark beetles, wood borers)	*Crypturgus*	X	Y	Y	1	C
	Dryocoetes	X	N	Y	1	C
			Y	Y	6	F
	Hylastes	NR	Y	Y	6	F
	Hylobius	X	Y	Y	1, 6	C, F
	Hylobius pinastri	NR	NS	Y	6	F
	Hylurgops	NR	Y	Y	6	F
		X	Y	N	1	C
	Ips	X	Y	N	1	C
	Orthotomicus	X	Y	N	1	C
	Pityogenes	NR	Y	Y	6	F
	Pityophthorus	X	N	Y	1	C
	Polygraphus	X	Y	Y	6	F
			N	Y	1	C
	Scolytus	X	N	Y	1	C
	Trypodendron	X	Y	Y	6	F
			Y	N	1	C
			Y	NS	6	F
Elateridae (typical click beetles)	*Ampedus*	NR	Y	Y	1	C
	Melanotus	NR	Y	NS	6	F
Latridiidae/Lathridiidae (minute brown scavenger beetles)	*Cartodere*	M	Y	Y	1	C
	Corticaria	M	Y	Y	1, 6	C, F
	Enicmus	NR	Y	NS	6	F
Leiodidae (round fungus beetles)	*Anistoma* *Agathidium*	M	Y	Y	6	F
Lycidae (net-winged beetles)	*Dictyoptera*	M	Y	Y	1	C
Melandryidae (false darkling beetles)	*Serropalpus*	X	Y	Y	1	C
	Xylita	X	Y	Y	1	C

(*continued overleaf*)

ECOLOGICAL ROLES OF WILDFIRE RESIDUALS

Table 4.1 (*continued*)

Family (common name)	Genus	Food category[a]	Timing (years after the fire)[b] 0–1	2–5	Source[c]	Location[d]
Melyridae (soft-winged flower beetles)	*Dasytes*	NR	Y	NS	7	F
Monotomidae (root-eating beetles)	*Rhizophagus*	NR	Y	Y	6	F
Ptiliidae (minute beetles)	*Acrotrichis*	M	Y	N	1, 6	C, F
	Ptiliolum	M	N	Y	1	C
Salpingidae (narrow-waisted bark beetles)	*Sphaeriestes*	P	N	Y	5	C
	Sphaeriestes stockmanni	NR	Y	Y	6	F
Scolytidae (bark and ambrosia beetles)	*Crypturgus*	Ph	Y	Y	5	C
	Dryocoetes	Ph	Y	Y	5	C
	Orthotomicus	Ph	NS	Y	5	C
	Polygraphus	Ph	Y	Y	5	C

[a]X, xylophagous; Ph, phloephagous; M, mycetophagous; P, predator; NR, not reported.
[b]Y, found; N, not found; NS, not sampled.
[c]1, Boulanger and Sirois (2007); 2, Evans (1964); 3, Muona and Rutanen (1994); 4, Saint-Germain *et al.* (2004a); 5, Saint-Germain *et al.* (2004c); 6, Toivanen and Kotiaho (2007); 7, Wikars (2001); 8, Wikars (2002).
[d]C, Canada; F, Finland; S, Sweden.

is attributed to progressive depletion of the nutrient-rich phloem and the cambial tissue of the dead trees. The rate of change depends on how soon the dead trees fall and on the decomposition rate of the fallen stems. Therefore, it is likely that the severity of the fire, the pre-fire tree species composition, and the type and frequency of post-fire disturbances such as windthrow will affect how quickly the second stage of the saproxylic communities begins.

Many insect species other than saproxylic beetles continue to arrive in burned areas after fires. Although some of these are arboreal species, many are epigeic species that inhabit the forest floor and feed on detritus, fungi, or other insects. These include carabid and staphylinid beetles (Muona and Rutanen 1994, Paquin 2008, Pohl *et al.* 2008). The populations of epigeic beetles also increase due to the abundance of dead wood on the forest floor from fallen snags and other debris (Boulanger and Sirois 2007). In addition, the burned forest-floor materials and decaying snag residuals begin to grow ascomycetous fungi, which are a major food source for many *fungivorous* insect species (also named *mycetophagous* because they feed on the fungal mycelia) (Wikars 1997). Common mycetophagous beetle species in Fennoscandia, for example, include lathridiids, cryptophagids, nitulids, and leiodids (Muona and Rutanen 1994). Many predatory carabids, such as *Sericoda, Harpalus, Bembidion, Pterostichus,* and *Elaphrus* species, also quickly occupy post-fire sites (Muona and Rutanen 1994, Cobb *et al.* 2007) to prey on other insects. As a result, the insect species diversity at these sites both increases and changes over time as the snag residuals decay and as debris falls to the ground (Grove 2002).

Implications of beetles in snag residuals

The post-fire saproxylic beetles have generally been considered pests from a timber supply point of view because they lower the quality of wood in two ways: by direct physical damage to the standing stems caused by boreholes and larval burrows, and by indirect damage in facilitating fungal infections that stain the wood (e.g., Richmond and Lejeune 1945, Ross 1960). However, their role in landscape-scale ecological processes and biodiversity is increasingly recognized (Saint-Germain *et al.* 2008). Some researchers have noted that most bark beetles and wood borers are also present in undisturbed forest stands, though at lower population levels (Sauvard 2004, Saint-Germain *et al.* 2007). Their populations reach epidemic levels only when triggered by periodic disturbances such as boreal wildfires that produce a large supply of ideal hosts in the form of stressed or dead trees. Therefore, from a biodiversity point of view, the survival of these species is a concern in areas where fires are frequently suppressed and the availability of snags and other deadwood is reduced (Langor *et al.* 2006). For example, Ahnlund and Lindhe (1992), quoted by Saint-Germain *et al.* (2004a), considered aggressive fire suppression as the reason for the near-extinction of several pyrophilous species in Fennoscandia.

As a result, some researchers recommend reintroducing fire in areas where wildfires have been suppressed effectively for long periods, so as to conserve the diversity of saproxylic insects (e.g., Grove 2002, Simila *et al.* 2002, Wikars 2002, Toivanen and Kotiaho 2007, Johansson *et al.* 2010). Some of these species are rare and have been *red-listed* – that is, proposed as candidate endangered or threatened species – in boreal Europe (e.g., Simila *et al.* 2002, Toivanen and Kotiaho 2007).

Although the ecological effects of insect invasions after boreal wildfires may last for decades, there are two notable immediate biotic consequences. First, these insects enhance post-fire nutrient cycling processes. The vast quantities of wood chips and insect frass also accelerate local nutrient cycling by increasing soil microbial activity (Cobb *et al.* 2010) and contribute to the revegetation of burned sites by restoring soil organic matter that was lost through burning. Because the influence of decomposing downed wood on the nutrient cycle in wildfire footprints is not significant, and downed wood releases nutrients slowly *in situ* (Laiho and Prescott 2004), the chips and frass may increase nutrient availability in the early years after a fire. Saproxylic beetles also create habitat for wood-decay fungi.

ECOLOGICAL ROLES OF WILDFIRE RESIDUALS

Galleries excavated by larvae weaken the wood, making it easier for fungi to penetrate (Rayner and Boddy 1988). Also, the adult beetles act as vectors for the spores of wood-decay fungi (Paine *et al.* 1997). These fungi increase the rate of decay, resulting in faster nutrient release from the dead wood (Edmonds and Eglitis 1989). The second biotic consequence, detailed in the next section, is that saproxylic insects initiate an influx of woodpeckers from outside the wildfire footprint and provide a steady source of food: the larvae of bark beetles and wood borers act as a short-term food source for the woodpeckers.

Colonization by woodpeckers

Once insects have colonized burned areas, it does not take long for bird life to return. Woodpeckers are the first to arrive, attracted by the pyrophilous insects. The influx of woodpeckers in pursuit of the beetle swarms that invade burned forests after a fire has drawn the interest of ecologists for decades. For example, Blackford (1955) noted that a woodpecker "task force" arrived soon after a 1945 fire and caused "extensive destruction" of the wood-boring beetles, and Axtell (1957) attributed the sudden influx of woodpeckers soon after large fires to the increased abundance of beetles. Since then, these woodpeckers and their occupancy of post-fire snag residuals in boreal forest landscapes have received considerable attention from researchers both in Fennoscandia (e.g., Angelstâm and Mikusinski 1994, Pakkala *et al.* 2002) and North America (e.g., Murphy and Lehnhausen 1998, Hoyt and Hannon 2002, Ibarzabal and Desmeules 2006, Saab *et al.* 2009, Nappi and Drapeau 2009). In North America, the woodpeckers attracted to burns include the black-backed (*Picoides arcticus*), American three-toed (*Picoides tridactylus*) (Figure 4.4), and hairy (*Picoides villosus*) species.

The natural range of these bark-probing woodpeckers spans the boreal forest: the black-backed and three-toed species are found primarily in coniferous forests, whereas the hairy woodpecker is widely distributed in a variety of habitats. These birds are not post-fire obligates (i.e., they do not require burned forests), but there are suggestions that burned forests provide better habitat (Hutto 1995, Bonnot *et al.* 2008) because of the unique combination of high food availability and low predation (Nappi and Drapeau 2009). When they inhabit unburned forests, these woodpeckers appear to prefer mature forests where they can use large snags for nesting and feed on the saproxylic beetles in dead and decaying stems (Tremblay *et al.* 2009). Here, we focus primarily on the three-toed and black-backed woodpeckers since they are considered to be fire-attracted species, whereas the hairy woodpecker is found in a variety of habitats.

The black-backed and three-toed woodpeckers are considered to be locally rare throughout most of their natural range, as has been reported for black-backed

(a)   (b)

Figure 4.4 Woodpeckers, and particularly the (a) black-backed (*Picoides arcticus*) and (b) American three-toed (*Picoides tridactylus*) woodpeckers, invade burned forests after a wild-fire to feed on saproxylic beetles. The three-toed woodpecker can be differentiated from the black-backed woodpecker, which has a solid black back, by the black and white bars on the center of its back. Males also have a yellow patch on the top of their head. (Reproduced with permission from A. Nappi, Government of Québec.)

woodpeckers in Alaska (USA) (Murphy and Lehnhausen 1998), and Québec (Canada) (Ibarzabal and Desmeules 2006). When a fire occurs, they move to and congregate at the burned site, causing a sudden and dramatic increase in local populations. For example, Fayt *et al.* (2005) noted that after a fire, local populations of black-backed woodpeckers can increase to 20 times the pre-fire numbers that existed in the surrounding forest.

Arrival of woodpeckers

The distance traveled by yearlings of many bird species depends on the suitability and occupancy of the habitat they encounter as they disperse (van Balen 1980). Given that woodpeckers have been observed in post-fire stands where they were not present before the fire (Blackford 1955, Murphy and Lehnhausen 1998), they clearly travel some distance from the surrounding forest to colonize recent burns. As part of a study documenting *irruptions* (sudden population increases) and movement between eastern Canada and the USA over many decades, Yunick

(1985) reported woodpecker movement of up to 570 km in one season; however, he thought this was a rare occurrence. Hoyt and Hannon (2002) found that when a fire occurred, both black-backed and three-toed woodpeckers were absent in mature forest stands within 50 km of the fire where they would otherwise occur. However, the birds continued to be present in mature stands beyond 75 km from recent fires, indicating a distance threshold somewhere between 50 km and 75 km. In northwestern USA, black-backed woodpeckers were reported as able to disperse less than 100 km and hairy woodpeckers as able to disperse only about 40 km (Pierson 2009). Therefore, it is likely that these woodpeckers will travel up to 50 km to arrive at burned areas.

Although it is unclear whether the woodpeckers arrive at fires as a result of some mechanism that attracts them to the fire itself or whether they actively follow the beetles, it is likely that they are searching for areas with increased food availability. Regardless of why they come or how far they travel, the woodpeckers arrive *en masse* following fires and are quick to find burns, moving in within a few months after the fire. In fact, we have noticed woodpecker occupation of burned sites within 2 months of a summer fire in northern Ontario (Canada). In Alaska (USA) woodpeckers occupied a June burn by September of the same year, and both black-backed and three-toed woodpeckers were abundant by the first winter (Murphy and Lehnhausen 1998). In both Québec (Nappi and Drapeau 2009) and Alberta (Hobson and Schieck 1999) (Canada) large numbers of black-backed woodpeckers were noted within 1 year after a fire. This rapid influx makes sense given the short-lived nature of the infestation by their major food source, the bark and wood-boring beetles; as discussed earlier in this chapter, these infestations persist for only a few years after a fire.

Feeding and breeding

Woodpeckers arrive at fires to feast on the beetle larvae that proliferate in the dead and dying trees. Their major food source is the larvae of wood-boring saproxylic beetles, with species preferences varying among woodpecker species. Black-backed and hairy woodpeckers prefer the wood borers, and especially the larvae of *Monochamus scutellatus*, the white-spotted sawyer beetle (Murphy and Lehnhausen 1998, Nappi and Drapeau 2009), whereas three-toed woodpeckers primarily eat bark beetle larvae (Murphy and Lehnhausen 1998, Fayt *et al.* 2005). However, these feeding patterns can change during a bird's life cycle. During the reproductive period, for example, three-toed woodpeckers may prefer larger sawyer beetle larvae due to their higher caloric content, regardless of the abundance of other beetle species (reviewed by Fayt *et al.* 2005). These woodpeckers can consume huge quantities of larvae; Baldwin (1968) reported that three-toed woodpeckers require approximately 3200 larvae per day to meet their caloric requirements, and Taylor and Barmore (1980) observed a three-toed woodpecker find and eat 35 sawyer beetles in as little as 5 minutes. These birds are therefore voracious feeders and must significantly moderate beetle populations in post-fire ecosystems.

Woodpecker use of snag residuals within a burn varies spatially and temporally, and is related to their feeding habits and the availability of specific beetles that occupy different types of snags. This variability is scale-dependent, observed at both the scale of a wildfire footprint and that of a single tree. It varies within wildfire footprints according to the fire severity, varies among trees based on their diameter, and fluctuates within trees based on variations in the availability of phloem and cambial tissue within a stem (Murphy and Lehnhausen 1998, Kotliar *et al.* 2008). In a snag residual, woodpeckers may prefer less-charred portions, possibly due to the higher quantity of phloem or higher moisture content in those locations and hence the higher availability of saproxylic insects (Nappi and Drapeau 2011). Generally, the best habitat for woodpeckers within fires is the areas of low fire severity, such as at the edges of burned areas. This is true for both the black-backed (Nappi and Drapeau 2011) and three-toed (Kotliar *et al.* 2008) species. Where fire severity is higher, such as in the interior of burns, no woodpeckers may be present (Murphy and Lehnhausen 1998). It is possible that their distribution within a wildfire footprint coincides with the distribution of the beetles they prey on.

Some species differences are apparent among woodpeckers in their use of snag residuals, in terms of both degree of burn and stem size. Three-toed woodpeckers appear to feed more on lightly to moderately burned trees, where they forage at shallow depths on bark beetles found only in the cambium. Black-backed woodpeckers feed on moderately to heavily burned trees, hairy woodpeckers feed on heavily burned trees, and both species feed in the sapwood (the portion of the tree stem or branch that is composed of phloem and cambium) and forage on beetles that bore into the wood (Murphy and Lehnhausen 1998). In their efforts to access the beetle larvae, woodpeckers create deep cavities on stems and remove much of the bark (Figure 4.5). It is generally agreed that they prefer larger-diameter snag residuals for foraging, likely because larger trees typically house higher densities of larvae, and therefore represent higher-quality foraging sites (Hobson and Schieck 1999, Nappi *et al.* 2003, Saint-Germain *et al.* 2004c, Nappi and Drapeau 2011). Because they prefer larger snag residuals, there may be a perception that woodpeckers have preferences among tree species. However, tree species could affect the preferred habitat for other reasons. Hoyt and Hannon (2002) indicated that three-toed woodpeckers preferred spruce snag residuals over jack pine (*Pinus banksiana*) snag residuals because the thick bark of the latter may make it more resistant to bark beetles. However, this was not the case for black-backed woodpeckers, which foraged on both spruce and jack pine snag residuals.

(a)   (b)

Figure 4.5   While foraging for beetle larvae among snag residuals, woodpeckers often (a) create large feeding cavities in and (b) remove large pieces of bark from burned stems.

In addition to foraging, woodpeckers also use standing dead trees for nesting, typically producing only one brood per year (Villard and Schieck 1996, Hoyt and Hannon 2002). Some species of snag residuals may be favored for nesting; for example, Nappi and Drapeau (2011) noted a preference for larger aspen (*Populus* spp.) trees that were present among the smaller black spruce (*Picea mariana*) stems. Nests consist of large cavities excavated from the burned stem, and have small, round entry holes. The young hatch in summer and disperse, but likely do not go far if food is abundant. These woodpeckers are non-migratory, so some will overwinter in the wildfire footprint and continue to forage on beetle larvae in subsequent years.

Departure of woodpeckers

Woodpeckers will remain in a burned area and use it as habitat only as long as their food source (beetles) and habitat (snag residuals) remain available. Generally, these woodpecker populations peak 2 years after a fire and then decline rapidly with the departure of juveniles (reviewed by Fayt *et al.* 2005). Both black-backed and three-toed woodpeckers begin to leave within a few years after the burn,

ECOLOGICAL ROLES OF WILDFIRE RESIDUALS

probably because after the beetle larvae become adults, usually within 2 years after a fire, the beetles leave in search of new burns. The birds seem to move on in pursuit (Figure 4.6), probably because their food source is depleted (Murphy and Lehnhausen 1998). The specific timing may vary based on their food preferences (i.e., types of beetles) and the severity of the burn, which affects the quantity of beetle habitat and therefore the abundance of beetles and the longevity of the population (Nappi *et al.* 2010). In general, woodpeckers vacate burned areas 3–10 years after a fire (Murphy and Lehnhausen 1998, Hoyt and Hannon 2002, Fayt 2003, Nappi and Drapeau 2009). As their food source diminishes and snag residuals begin to fall, the birds move on to new wildfires – locally or regionally if such areas are available – or redistribute into forested areas, where they likely remain at low abundance until new wildfires that can support larger populations become available within their dispersal range.

Consequences of woodpecker occupation of snag residuals

In foraging for beetles, woodpeckers remove the bark of the trees to expose the beetle galleries and create large cavities in the wood. This activity likely accelerates the breakdown of the snag residuals, and provides entry points for fungi and other decomposers. Also, the nesting cavities left behind in standing snag residuals (Figure 4.7) become useful to secondary cavity nesters, making it easier for them to colonize post-fire areas. These later arrivals are known as secondary cavity nesters because they cannot excavate trees on their own but can opportunistically occupy the nests vacated by the woodpeckers, which are the primary cavity nesters (Nappi *et al.* 2004). For example, eastern bluebirds (*Sialia sialis*) and tree swallows (*Tachycineta bicolor*) have been found to use the nesting cavities produced by black-backed woodpeckers in the second and third years after a fire (Hannon and Drapeau 2005). Many other bird species and mammals also use existing cavities, and most of these species have been found to use the cavities produced by the large-bodied woodpeckers (Gentry and Vierling 2008). In fact, woodpeckers are considered a keystone species in boreal forest ecosystems because a spectrum of species, including breeding bats (*Myotis* spp.), owls, and chickadees (*Poecile* spp.) use the tree cavities they excavate (Thompson and Anglestâm 1999).

The need to sustain populations of woodpeckers in the interest of regional biodiversity has been recognized in some boreal forest landscapes, particularly in Fennoscandia (e.g., Pakkala *et al.* 2002, Saab *et al.* 2005). This is based on the premise that bark-probing woodpeckers depend on the proliferation of bark and wood-boring beetles after recent fires, and thus require a steady supply of wildfires of varying levels of severity. As noted previously, concerns have been raised about the challenges of maintaining saproxylic beetle populations in regions such as Fennoscandia given modern fire-suppression efforts. Aggressive and effective

ECOLOGICAL ROLES OF WILDFIRE RESIDUALS

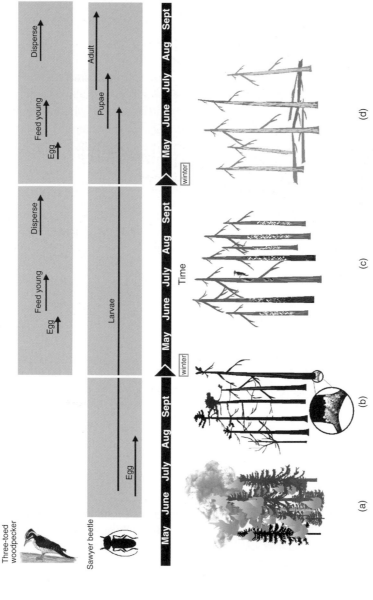

Figure 4.6 The life cycle of the American three-toed woodpecker (*Picoides tridactylus*) is synchronized with that of the white-spotted sawyer beetle (*Monochamus scutellatus*) in terms of the timing of their use of burned areas. (a) The onset of a wildfire. (b) Adult beetles arrive within a couple of months of the fire to breed and lay eggs, and feeding by their larvae is indicated by the massive quantities of wood residue that appear around infested stems. Woodpeckers arrive and begin nesting. (c) The woodpeckers continue to feed on beetle larvae and produce a second brood. (d) The beetles pupate and depart as adults, the woodpeckers also depart, and the snag residuals begin to fall.

Figure 4.7 A woodpecker nesting cavity in a snag residual during the second year after a fire. After the woodpeckers depart, these nesting cavities become useful for secondary cavity nesters for a decade or so after the fire.

fire suppression efforts that diminish the supply of such fires at broader scales can adversely affect woodpecker populations.

A regional-scale hypothesis for relocation of local populations is that if fires create a shifting spatial mosaic (i.e., if wildfire footprints occur at regular temporal intervals somewhere in the landscape), with new fires occurring within a certain distance of woodpecker populations, then the older recent burns are sources for dispersal to new wildfires, and the new burned areas become the locations of convergence. In this scenario, mature unburned forests may play a less significant role in sustaining long-term woodpecker populations. Alternatively, if fires do not occur at a sufficient frequency or if they occur beyond the relocation distance of woodpecker populations, then older recent burns may still initially serve as dispersal sources and the unburned mature forests in proximity to these sources may serve as destinations. The latter may subsequently become major sources for dispersal when fires eventually occur within the relocation distance of the woodpeckers (e.g., 50 km). However, if fires continue to be absent within the maximum relocation distance, then the woodpeckers may have to migrate over longer distances searching for burned areas, possibly even beyond the regional scale.

As Yunick (1985) suggested, local black-backed woodpecker populations may remain sedentary until a large-scale forest disturbance occurs nearby, and their populations may therefore undergo periods of subsistence before resources increase sufficiently to permit a population increase. This may be the reason for observations that these woodpeckers are locally rare within their range both in boreal forests (Murphy and Lehnhausen 1998) and other fire-dominated forests (Hutto 1995).

Other disturbances that generate snags in large numbers can also provide suitable habitat. For example, Bonnot et al. (2008) pointed out that stands damaged by the mountain pine beetle (Dendroctonus ponderosae, Coleoptera: Curculionidae) are equivalent to post-fire habitats for black-backed woodpeckers, and Pakkala et al. (2002) mentioned the value of windthrown trees for three-toed woodpeckers. Although the disturbance required to permit population increases by woodpeckers does not have to be a wildfire, it is clear that, in the absence of frequent burns, the periodicity and extent of these non-fire disturbances may be too low to trigger increases in woodpecker populations.

Occupation by other bird species

Although other bird species have not been studied as extensively as woodpeckers, many are known to occupy burned areas. Songbird assemblages after a boreal wildfire have been documented as distinctly different from those in undisturbed forest (e.g., Morissette et al. 2002). Many birds forage in burned areas after boreal wildfires; they include timber gleaners, tree foliage searchers, flycatchers, ground and brush foragers, and raptors (Figure 4.8). The appearance of some species in wildfire footprints depends on the time since the fire. It may coincide with the recruitment of new snags as live tree residuals die and snag residuals fall, with the emergence of new understory vegetation from the seed bank or from vegetative organs that survived the fire, and with the arrival of prey species. For example, seed eaters such as crossbills (Loxia spp.) and pine siskins (Carduelis pinus) are attracted to early post-fire stands, where they feed on coniferous seeds released by fire and become abundant in the first and second years after a fire (Hutto 1995); in fact, pine siskins are known to arrive within weeks of a fire (Taylor and Barmore 1980). Raptors such as the northern hawk owl (Sumia ulula) are attracted by the rodents that increase in abundance among the snag residuals, and their peak nesting period occurs within 3–5 years after a fire (Hannah and Hoyt 2004).

Populations of secondary cavity nesters increase over time as the initial inhabitants of the cavities (the woodpeckers) abandon their nests and depart around 2 years after a fire (Taylor and Barmore 1980). Although we did not find any scientific literature that specifically reported usage durations in boreal forests,

Chapter 4   141

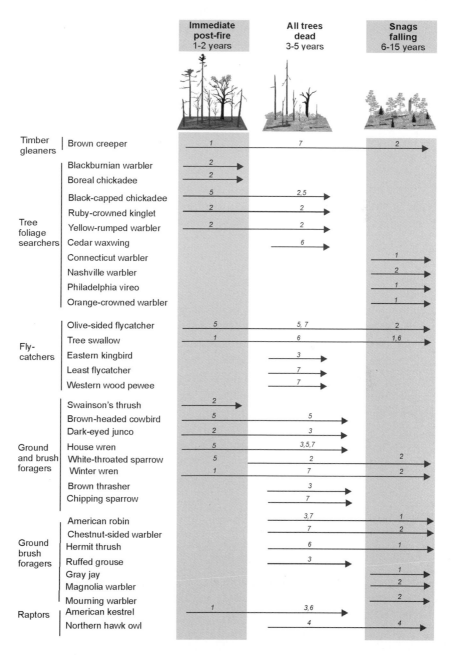

Figure 4.8 There appears to be a regular pattern in arrival time of bird species guilds other than woodpeckers for the use of post-fire stands in boreal North American forests. Sources: guild designations follow Bock and Lynch 1970, cited by Haney *et al.* 2008. 1, Hobson and Schieck 1999; 2, Haney *et al.* 2008; 3, Niemi 1978; 4, Hannah and Hoyt 2004; 5, van Wilgenburg and Hobson 2008; 6, Imbeau *et al.* 1999; 7, Morissette *et al.* 2002.

woodpecker cavities seem to be used for up to a decade after a fire. For example, in a ponderosa pine (*Pinus ponderosa*) system in Idaho, Saab *et al.* (2004) found that the re-use of woodpecker cavities by other species increased for up to 4 years following a fire and then declined during the rest of the first decade. They also indicated that the interspersion of cavity trees within live tree residuals and residual patches promotes cavity re-use.

Even though ample information is available for birds that inhabit recent burns in other fire-driven ecosystems, such as the reviews by Hutto (1995) and Saab *et al.* (2005), little knowledge is available about the population dynamics of birds other than woodpeckers in burned boreal forests. It is thus uncertain whether the presence of these bird species is merely facultative (i.e., they can just as easily occupy other habitats) or obligatory (i.e., their presence in post-fire habitat is not a coincidence because burned areas represent an essential part of their habitat).

Forest regeneration

In addition to providing habitat, post-fire residual vegetation in a wildfire footprint provides a source of propagules to regenerate the next forest. Post-fire regeneration of burned areas occurs from *in situ* propagules, such as seeds stored in the canopy or plant parts that survived the fire and through an influx of wind-dispersed seeds or spores from unburned areas. These plant regeneration strategies were grouped by Rowe (1983) into categories: plants that *evade* fire, which are killed by the fire but whose seeds are preserved; those that *endure* fire, and regenerate from persistent or buried parts that escape lethal heat; and those that *invade*, which have wind-dispersed propagules that are blown into the burned area. All three strategies are relevant for post-fire residual vegetation – the first two involve *in situ* residuals in the wildfire footprint and the third involves unburned areas that lie either within the wildfire footprint (as residual patches) or adjacent to the wildfire footprint (as remnant stands). In this section, we focus on *in situ* residual vegetation and its role in regenerating the burned area; we do not consider the soil seed bank as post-fire residual vegetation, and do not address it here.

Regeneration by serotinous cones

Two major boreal forest species – jack pine and black spruce – are evaders. They are abundant in the post-fire ecosystem as snag residuals, and play a large role in re-establishing the next forest through seed stored in the canopy. Jack pine cones are *serotinous*, which means "late to open", and require heat to open and shed their seed. The bonding material that holds the cone scales together melts when

Figure 4.9   (a) Jack pine (*Pinus banksiana*) has (b) serotinous cones that are opened by the heat of a fire. As a result (c), their seedlings become established immediately after a fire, and quickly colonize the burned area. These seedlings have emerged barely 2 months after a boreal wildfire.

subjected to high temperatures, such as during a fire (Beaufait 1960), allowing the cones to open and release seeds. Also, seeds in the cones are insulated and can survive short bouts of exposure to flame (Eyre 1938). Jack pine can produce cones as early as only 3–5 years of age; in addition, it has annual cone crops and retains large numbers of cones in its crown (Eyre and LeBarron 1944). Cones can accumulate on the trees for up to 25 years, and the seeds remain viable for more than 20 years (Cayford and McRae 1983), so regardless of when a fire occurs, it is likely that abundant viable seeds are available locally. These provide an *in situ* source of propagules for regeneration (Figure 4.9). When fire intensity is low, jack pine cones may not open, and subsequent regeneration will be sparse (Noël 2001, cited in Jayen *et al.* 2006).

Black spruce has semi-serotinous cones, which remain on the tree and remain closed beyond their maturation period, and can store seed for several years until they are opened by heat (Viereck 1983). The cones are clustered near the top of tree crown, where they are less likely to be damaged by the heat of the fire (Lutz 1956, Ahlgren 1974). As a result, if their crowns are not completely scorched and the fire occurs in a seed year, both live and dead trees will provide a seed source during the first few years after a fire (Figure 4.10).

(a)

(b)   (c)

Figure 4.10   (a) As long as the cones are not completely scorched, black spruce (*Picea mariana*), which has semi-serotinous cones, will disperse seed (b) for a few years after a fire, providing a local seed source for regeneration. (c) This seedling emerged shortly after a boreal wildfire and has survived its first 2 years.

Snag residuals are a critical source of seed that will regenerate the post-fire forest. Eyre (1938) reported that jack pine seed fall began almost immediately after a fire, and that most of the seed had fallen within 4 days, leading to rapid regeneration. Jayen *et al.* (2006) found that the density and basal area of cone-bearing snag residuals positively influenced the post-fire regeneration of jack pine and black spruce seedlings 6 and 7 years after a fire, respectively, in Québec (Canada). The consensus is that recruitment of jack pine and spruce seedlings peaks 1–7 years after a fire (Ahlgren 1959, Greene and Johnson 1999, Greene *et al.* 2004, Johnstone *et al.* 2004); spruce germination is sometimes delayed by 1–2 years, possibly because the ash is initially toxic to spruce (Ahlgren 1959). Others have also indicated that charred surfaces may detrimentally affect post-fire regeneration. Foster (1985) indicated that black spruce seedlings could not survive the heat and exposure of the post-fire environment. Sirois (1993) reported lower germination rates on severely burned seedbeds, which he attributed to the development of a hydrophobic crust, as well as to higher temperatures and lower water retention. Despite this possible initial delay, prolific post-fire regeneration of jack pine and black spruce is common in wildfire footprints.

As we discussed in Chapter 2, snag residuals remain standing for long periods – sometimes up to 20 years for jack pine. Therefore, they could remain a significant source of viable seeds, especially during the period when seedbeds are most receptive after a fire (typically less than 3–4 years). In addition to providing a ready source of seed, snag residuals may benefit the regeneration of some species by shading and ameliorating the local environment (Figure 4.11).

Figure 4.11   Snag residuals, shown here 1 year after a wildfire in a black spruce (*Picea mariana*) stand, can significantly shade the forest floor, influencing subsequent regeneration. (Reproduced with permission from W.C. Parker, Ontario Ministry of Natural Resources.)

ECOLOGICAL ROLES OF WILDFIRE RESIDUALS

Windthrown snag residuals may also benefit regeneration. In northern Sweden, shade from snag and live tree residuals initially favored Norway spruce (*Picea abies*) regeneration, but when the standing wood fell and the area opened up, providing more light, Scots pine (*Pinus sylvestris*) began to regenerate (Engelmark 1993). Also, clumps of windthrown timber may protect regenerating seedlings from ungulate browsing. This was the case for aspen (*Populus tremula*) and willow (*Salix caprea*) in Sweden (de Chantal and Granström 2007), where larger snag residuals with larger crowns provided protection for seedlings against moose and deer for longer periods than smaller clumps because they broke down more slowly. This was also true for trembling aspen suckers in Yellowstone, where groups of 2–10 fallen conifers – referred to as "jackstraw piles" – provided protection against elk browsing (Ripple and Larson Ripple and Larsen, 2001). The best protection was provided by piles >8 m tall formed by trees with many branches. Woody debris has been shown to be important for the survival of newly regenerated seedlings in burned areas, as it creates a microenvironment with increased moisture and shade (Jayen *et al.* 2006).

Regeneration from surviving underground structures

In addition to the canopy-stored seed that regenerates the forest in burned areas, vegetative reproduction from root suckers, basal stem sprouts, rhizomes, tubers, and bulbs is a primary source of plant re-establishment in the boreal ecosystem after a fire. This is particularly true for shrubs (Table 4.2). Sirois (1995) indicated that most species that regenerate in burned black spruce–lichen woodlands in Québec (Canada) are endurers that sprout from vegetative parts; these include Labrador tea (*Ledum groenlandicum*), blueberry (*Vaccinium* spp.), and hazel (*Corylus* spp.). Foster (1985) indicated that in Labrador (Canada) the primary mode of post-fire regeneration was sprouting or suckering from underground organs, and provided examples of common species that regenerated in this manner: raspberry (*Rubus* spp.), hazel, blueberry, Labrador tea, and horsetails (*Equisetum* spp.). Ahlgren (1959) and Heinselman (1981) provided a similar list of species that reproduced vegetatively after a fire, with root suckers being particularly important for aspen (*Populus tremuloides*), and root or stem sprouts important for birch (*Betula* spp.), alder (*Alnus* spp.), hazel, and blueberry (Figure 4.12). In a spruce–lichen forest, trees and lichens were killed by a severe fire (there were no survivors), but a shrub layer composed of birch and Labrador tea survived through persistent rootstocks, whereas blueberry and crowberry (*Empetrum* spp.) decreased in abundance, likely due to their shallower root systems (Auclair 1985).

Viereck (1983) noted that less-intense fires that kill aboveground plant parts and char and kill the moss layers will not kill the belowground parts of many

Table 4.2   Post-fire residual vegetation is a source of propagules for regeneration of the burned area. Examples given (adapted from Heinselman 1981, Rowe 1983) indicate how residuals contribute to regenerating the burned area for various plant life forms that rely on different regeneration strategies.

Residual source	Life form and species	Primary source of propagules following fire disturbance	Regeneration strategy with respect to fire (*sensu* Rowe 1983)
Snags – a source of canopy-stored seed	**Trees** Jack pine (*Pinus banksiana*) Black spruce (*Picea mariana*)	Serotinous cones Semi-serotinous cones	Evasion
Underground structures – a source of soil-stored propagules	**Trees** Aspen (*Populus* spp.) Birch (*Betula* spp.) Willow (*Salix* spp.)	Root suckers Stem sprouts Root or stem sprouts	Endurance
	Tall shrubs Alder (*Alnus* spp.) Currant (*Ribes* spp.) Dogwood (*Cornus* spp.) Hazel (*Corylus* spp.) Raspberry (*Rubus* spp.) Viburnum (*Viburnum* spp.)	Stem sprouts Stem sprouts Stem sprouts Stem sprouts Root suckers, sprouts Root collar sprouts	
	Low shrubs Blueberry (*Vaccinium* spp.) Labrador tea (*Ledum groenlandicum*)	Rhizomes Rhizomes	
	Other plants Horsetails (*Equisetum* spp.)	Rhizoids	

shrubs and herbs, which can then regenerate from their rhizomes (e.g., blueberry and Labrador tea) or sprouts [e.g., willow (*Salix* spp.), birch, rose (*Rosa* spp.), and raspberry]. Horsetails and hair-cap moss (*Polytrichum* spp.) have rhizoids that grow into the mineral soil and that will survive even intense burns that remove the organic matter (Viereck 1983). Plant propagules can survive as long as soil temperature does not exceed the lethal level of >60 °C (as discussed in Chapter 2). These early-establishing shrubs and mosses are key to re-establishing plant cover in post-fire systems. In addition, they have been shown to provide browse for ungulates that feed in burned areas and use the surrounding unburned areas for cover (Hebblewhite *et al.* 2009).

(a)   (b)

(c)   (d)

Figure 4.12 Burned forests are rapidly colonized by trees and shrubs that sprout from underground structures such as roots and stems. Common species include (a) alder (*Alnus* spp.), (b) willow (*Salix* spp.), (c) birch (*Betula* spp.), and (d) aspen (*Populus* spp.). These photographs were taken 2 months after a boreal wildfire.

In the first few years after a fire, the burned area is recolonized by *in situ* species from surviving plant parts that occur both above and below the ground. These plants play a role in re-establishing both the tree cover and the understory of the new forest, and contribute to ecosystem recovery through nutrient cycling and retention and by providing habitat for other species that can subsequently recolonize the burned area. In addition to providing forage for ungulates, these species provide cover and food for small mammals and amphibians, and provide perches, food, and foraging locations for birds that prefer open habitats.

Roles of the residual patches

The aftermath of most boreal wildfires, regardless of intensity, is a heterogeneous spatial mosaic created by varying degrees of fire severity. Within the spatial mosaic, some areas may escape the fire ("fire skips") for various reasons (as we explained in

Chapter 2), show no signs of burning, and are true relicts of the pre-fire ecosystem. Such patches are distinct from clusters of live tree residuals that survived the fire, at least temporarily, where the forest floor, undergrowth, and part of the trees may have been burned. The unburned vegetation occurs in patches of various sizes and at various scales within the overall wildfire footprint; these range from a few square meters of unburned forest floor to several hectares of intact forest. These residual patches not only represent the assemblages of pre-fire plant and animal species, but also retain the original forest floor, soil, microorganisms, and micro-climatic conditions. Therefore, residual patches in wildfire footprints are regarded as "biological legacies" in forest landscapes (DeLong and Kessler 2000).

Here, we examine the ecological roles of these residual patches in boreal forest landscapes, with respect to minimizing the adverse effects of fire on animal habitat in the immediate term, during and soon after fire; providing ecological processes that facilitate the recovery of the burned ecosystems in the intermediate term; and contributing to the heterogeneity of boreal forest landscapes long after the burned area has recovered from fire.

Providing temporary shelter

It has been long suggested that the residual patches that escape the fire provide immediate refuge for many animal species at multiple scales (Krefting and Ahlgren 1974). There are many anecdotal references to highly mobile animals that seek refuge in unburned vegetation patches during a wildfire (Chandler *et al.* 1983). However, direct empirical evidence for this phenomenon is sparse in the literature on boreal wildfires. In one rare mention, Buech *et al.* (1977) reviewed the effect of wildfire on small mammal populations in northeastern Minnesota, and cited several references to residual patches that provided shelter both during and immediately after fire. They stated that small mammals (e.g., deer mice, *Peromyscus manicula-tus*) escaped the fire by moving into residual patches during the fire, and ascribe the survival of their populations to the existence of patches of unburned habitat.

There is, moreover, some indirect evidence from boreal wildfires that residual patches provide immediate shelter for wildlife. For example, Ellison (1975) reported the occurrence of spruce grouse (*Canachites canadensis*), a species that prefers closed-canopy coniferous forest, after a wildfire in Alaska (USA). Most of the male spruce grouse in his study were found in the residual patches during the summer after the fire. Other researchers have reported that food availability remains high for some species in the residual patches after a severe wildfire, and that these areas consequently provide temporary habitat for those animals. After severe fires in Alaska, residual patches provided browse and cover for moose (*Alces alces*) and the occurrence of moose in these patches was more than four times the levels in burned areas (Gasaway and Dubois 1985, Nelson *et al.* 2008).

Evidence for the value of residual patches from non-boreal wildfires includes a review of studies by Pilliod *et al.* (2003) of amphibians, mostly in Australia and western North America. Their review suggests general trends that may also be applicable in boreal forest landscapes. For example, they assert that the heterogeneity of a fire (i.e., the mix of burned and unburned areas) is crucial for amphibian survival, which varied among species, populations, and life stages. Generally, unburned riparian areas and forest remnants acted as refugia that contributed to population survival by providing moist areas and shelter from exposure, thereby facilitating thermal regulation (i.e., the maintenance of a suitable body temperature) by salamanders and frogs. A series of studies by Hossack and colleagues (Hossack *et al.* 2006, Hossack and Corn 2007) in Montana (USA) showed that toads need moist areas and shelter from exposure to permit thermal regulation and survival after a wildfire. This finding agrees with the reports by Popescu and Hunter (2011), who found that patches of unburned vegetation in Maine (USA) provided shelter for wood frogs (*Lithobates sylvaticus*) during wildfires.

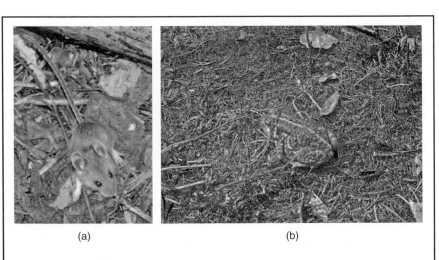

(a)   (b)

Residual patches provide refuges for mobile animals during and immediately after the fire, from which they may recolonize the burned area in time. Species known to seek shelter in residual patches within wildfires include small mammals, such as (a) the deer mouse (*Peromyscus maniculatus*) and (b) amphibians, such as the American toad (*Anaxyrus americanus*).

(Reproduced with permission from: (a) J. Bowman; (b) R. Routledge, Ontario Ministry of Natural Resources.)

Although direct empirical evidence is not available, it is highly probable that more mobile species may seek shelter in residual patches. If this is the case, it is plausible that animals preferentially search for a particular type of shelter if the residual patches result from specific biotic and abiotic conditions (e.g., wetlands, wind shelter) that are attractive to them. The role of residual patches in providing temporary shelter becomes apparent during the subsequent stages of post-fire changes within a wildfire footprint. For example, Fisher and Wilkinson (2005) cite many others in their review of post-fire changes to wildlife species to indicate that, over time, species (e.g., deer mice) migrate from residual patches into adjacent burned areas. Paragi *et al.* (1996) describe the same process with American marten (*Martes americana*), for which recent burns act as a sink for their populations. These studies suggest that animals may have moved into, or survived within, residual patches during and immediately after wildfires. It is unfortunate that empirical evidence is not available to support or refute the hypothesis that residual patches act as population sinks for mobile animal species during and after boreal wildfires and, furthermore, that the effects of the sizes, shapes, and locations of residual patches on this flow of organisms remain unknown.

Supplementing the recovery

Burned areas created by boreal wildfires may take many decades to re-establish forest cover and regain the pre-burn species assembly. Residual patches, due to their undisturbed nature and juxtaposition to the burned area, appear to play important roles in this recovery process.

Providing extended refuges and habitats

Residual patches continue to offer refuge to species beyond the temporary shelter provided during or soon after a fire. This service can last until suitable habitat is re-established in the burned areas and the new populations have reached pre-burn levels. The process is called "lifeboating", a term used by Franklin *et al.* (2000) to signify the role of residual patches. Lifeboating in residual patches is vital for the recovery of some species, such as epigeic beetles. Ground beetles and rove beetles survive in patches of unburned vegetation after boreal wildfires (Gandhi 1999). In fact, Gandhi *et al.* (2001) found 12 rare carabid beetle species only in the residual patches even two to three decades after a fire. Their survival in unburned areas of wetland and older forest is important for survival of the local population because these species are very sensitive to changes in their habitat (Gandhi *et al.* 2001, Cobb *et al.* 2007).

Fisher and Wilkinson (2005) and Nelson *et al.* (2008) summarized the findings from numerous studies in which mammals such as fisher (*Martes pennanti*) and marten found refuge in undisturbed residual patches after wildfires. They proposed that the unburned areas continue to provide more suitable habitat than the recovering burned area because they retain the pre-burn forest structure and composition. For small mammals, the moist microclimate, shelter from exposure, and a food supply in the form of fungi, lichens, and invertebrates within the residual patches, provide the habitat attributes necessary to form a lifeboat until the burned area develops sufficiently. A continuous supply of food such as lichens and fungi is essential for the survival of some species, including arboreal mammals such as the northern flying squirrel (*Glaucomys sabrinus*).

The continuous habitat supplied by residual patches is important even for highly mobile species such as birds. In one case, a "unique" bird species composition was found in residual patches that was typical of the bird species composition in old unburned forests in northern Minnesota. The birds in these patches included golden-crowned kinglet (*Regulus satrapa*), black-and-white warbler (*Mniotilta varia*), Nashville warbler (*Vermivora ruficapilla*), northern parula (*Parula americana*), magnolia warbler (*Dendroica magnolia*), Blackburnian warbler (*Dendroica fusca*), scarlet tanager (*Piranga olivacea*), and rose-breasted grosbeak (*Pheucticus ludovicianus*) (Niemi 1978). The size of residual patches may also affect their ecological role, since larger patches provide more habitat opportunities. For example, Schieck and Hobson (2000) found that bird communities in large residual patches were similar to those in old unburned forest in the proximity.

Some species that may thrive in residual patches because of the shelter provided, may nonetheless use the burned area to feed in. For example, Stuart-Smith *et al.* (2002) hypothesized that some bird species use the combination of unburned and burned areas resulting from boreal wildfires, nesting in the residual patches and feeding in the burned area. They mentioned the western wood-pewee (*Contopus sordidulus*) and the winter wren (*Troglodytes troglodytes*) as examples of such species. They also reported that 5–6 years after a fire in boreal Alberta (Canada) the residual patches had more bird species than the burned areas and even more than in areas of old forest outside the wildfire footprint. This was the case because the residual patches housed more aerial foragers than the undisturbed forest beyond the wildfire footprint, as well as more foliage gleaners than the burned area. Interestingly, some species were found only within the residual patches: the blueheaded vireo (*Vireo solitarius*), ovenbird (*Seiurus aurocapilla*), yellow-rumped warbler (*Dendroica coronata*), and yellow-bellied sapsucker (*Sphyrapicus varius*).

ECOLOGICAL ROLES OF WILDFIRE RESIDUALS

(a)   (c)

Residual patches within wildfire footprints may allow many organisms to survive by providing habitat (i.e., shelter and food) until the burned area recovers, a concept commonly referred to as "lifeboating". Small mammals such as (a) the American marten (*Martes americana*) and (b) ground-dwelling beetles (top, *Pterostichus* sp.; bottom, *Bembidion* sp.) are examples of species that can persist in these patches until the burned area recovers sufficiently to support them.

Note that the images of beetles are not life-size or consistently scaled.

(Reproduced with permission from: (a) Ontario Ministry of Natural Resources; (b) D. Ingram, Island Nature; (c) D. Maddison, Oregon State University.)

ECOLOGICAL ROLES OF WILDFIRE RESIDUALS

Providing an in situ *source of biota*

Recovery of a burned forest system requires an inflow of animals, both adult and juvenile, and plant propagules to replace the individuals that have been lost. Though the most obvious source of this influx is the unburned forest that surrounds wildfire footprints and provides an *ex situ* supply of biotic material, residual patches within the wildfire footprint also play a key role in this recovery process by providing an *in situ* source. Residual patches within wildfire footprints may be even more important because of their proximity to suitable sites within the burn and their ability to serve as a ready source of new individuals to repopulate the burned area (Hooven 1969, Buech *et al.* 1977). Residual patches act as a source and the burned area acts as a sink during the early stages of recolonization of the wildfire footprint. This process has been noted with small mammals such as deer mice (Fisher and Wilkinson 2005) and marten (Paragi *et al.* 1996) after boreal wildfires.

Gandhi (1999) proposed that, after a boreal wildfire, large residual patches are suitable places for adult carabid beetles to survive the fire and provide a critical breeding area for their reproduction. These patches then become an important source for recolonization of the recovering burned areas by epigeic beetles. Litter-dwelling insects play a vital role in breaking down coarse woody debris, and hence promote nutrient cycling, and their ready recolonization of the burned area is an important step for recovery of the ecosystem. Similarly, the rapid recovery of spider populations in the burned area enhances soil litter dynamics and provides a food source for overwintering birds, and spider populations are supplemented by short-distance dispersal from nearby residual patches (Buddle *et al.* 2000).

Repopulation of a burned area by species that have relatively short dispersal distances is also aided by the presence of residual patches within the wildfire footprint. This is likely to be the case for three species groups discussed below, which are functionally essential to the dynamics of boreal forest ecosystems: lichens, mosses, and ectomycorrhizae. We also briefly discuss herbaceous plants with limited seed-dispersal distances.

Lichens grow in open areas and are important to trees because they reduce soil moisture stress and fix nitrogen, and despite their soil insulating properties that may decrease the rate of organic matter decomposition, lichens generally improve tree growth (Bonan and Shugart 1989). Lichens, and particularly *Cladonia* and *Cladina* species, are also an important source of winter food for boreal caribou (*Rangifer tarandus caribou*), especially during late winter. Typically, lichens are destroyed during a boreal wildfire and their regrowth in a burned area can take several decades (Morneault and Payette 1989). Soredia, which are small vegetative propagules that disperse the fungal and algal components of lichen simultaneously, are considered the primary mode of recolonization (Johansson *et al.* 2006), but surviving lichen fragments also play a role for some species (Webb 1998). As

(a)

Sustainability of (a) woodland caribou (*Rangifer tarandus caribou*) habitat in North American boreal forest landscapes has become the focus of forest policy debate, spawning many research studies. One topic that has been widely researched is the supply of caribou food sources. (b) Lichens, especially the *Cladonia* and *Cladina* species, which are fruticose (shrubby) lichens, form a major food source for caribou, in some areas comprising as much as 90% of their winter forage.

(b)

Fires greatly decrease caribou lichen forage in the short term. However, fire is considered necessary to maintain optimal long-term lichen resources. This is because in the absence of fires, lichens are likely to be outcompeted by moss ground cover. Lichens become re-established in burned areas within 10–15 years after a fire and can form the dominant vegetation cover in open boreal pine and spruce forests from 20 to more than 100 years after a fire. The role of residual patches in the timing and extent of lichen re-establishment in burned areas remains one of the unknowns.

(Photo (a): G. Racey. Reproduced with permission of Ontario Ministry of Natural Resources.)

Johansson (2008) stated in a review of the effects of disturbance on lichens in boreal and near-boreal forests, lichens persist in residual patches. The recovery of lichen within a wildfire footprint depends on the arrival of diaspores from nearby sources (Morneault and Payette 1989). The dispersal of fungal symbiote spores is essential in this process, but these spores appear not to spread over great distances; in fact, their dispersal range may be limited to less than 100 m (Dettki *et al.* 2000). Therefore, it is possible that, given their close proximity to the burned areas, residual patches will be a key source for the influx of lichen propagules into burned areas. The role of residual patches in the timing and extent

ECOLOGICAL ROLES OF WILDFIRE RESIDUALS

of lichen re-establishment in burned areas remains one of the unknowns in the understanding of boreal fire residuals.

Bryophytes (mosses) replace lichens as the dominant ground cover in shaded or closed-canopy conditions (Bonan and Shugart 1989) and play several roles in the post-fire ecosystem: they stabilize charred surfaces, decrease soil erosion, and possibly decrease nutrient losses from burned sites, while increasing site moisture-holding capacity (Foster 1985, Ryoma and Laaka-Lindberg 2005). Although light windborne spores are the primary form of moss recolonization of burned areas, these spores are sufficiently fragile that their dispersal distances are limited. Therefore, adjacent residual patches may be the most important source of moss propagules for the burned area. In fact, most post-fire recolonization of mosses is thought to result from aerial dispersal of spores from adjacent unburned areas or from residual patches within the wildfire footprint (Ryoma and Laaka-Lindberg 2005, Bradbury 2006). Mosses that are exposed to fire but not killed (an example of micro-scale residual patches) can re-establish from surviving tissue in less-burned substrate. They will produce sporophytes and vegetative diaspores that disseminate rapidly, indicating that heterogeneity in the wildfire footprint is very important for the recovery of these mosses (Bradbury 2006).

Mosses and lichens are an important component of the ground cover that greatly influences the overall course of succession (Foster 1985). For example, hair-cap moss (*Polytrichum* spp.) provides an ideal seedbed for black spruce, jack pine, and trembling aspen, and its abundance influences seedling abundance in the first few years after a fire (Jayen *et al.* 2006). Ryoma and Laaka-Lindberg (2005) noted that, as vascular plants become established, the early colonizers are replaced by feathermoss (*Pleurozium schreberi*; Figure 4.13a).

Ectomycorrhizae (Figure 4.13b) are crucial for boreal forest ecosystems because their obligatory symbiosis with tree species, particularly conifers, assists them in absorbing nutrients and water more efficiently (Visser and Parkinson 1999), and they also assist carbon and nutrient cycling processes in the forest soil (Treseder *et al.* 2004). These fungi also die during fires, and few survive in the burned area due to the death of their host trees and the heating of the organic layer in which they live (Dahlberg 2002). Recolonization via spores that blow into the burned area from adjacent unburned areas is especially important in areas that have experienced a high-intensity fire (Visser 1995, Jonsson *et al.* 1999), as fungal mycelia can spread only over short distances and do so slowly (e.g., spread distance <30 m and spread rate <1 m annually; Dahlberg 2002). Fortunately, residual patches harbor ectomycorrhizal populations from which spores can be disseminated into the burned area, where they help the succeeding trees to recolonize the site and to re-establish the former symbiotic relationship in the recovering ecosystem (Dahlberg *et al.* 2001). Also, small mammals that move from

ECOLOGICAL ROLES OF WILDFIRE RESIDUALS

(a)   (b)

Figure 4.13   (a) Mosses and (b) ectomycorrhizae are critical components of boreal forest ecosystems. They support the re-establishment of the post-fire system by increasing, for example, the moisture-holding capacity of the seedbed (mosses), thereby facilitating regeneration, and by improving water uptake and nutrient cycling by tree species (ectomycorrhizae).

residual patches into adjacent burned areas may help to spread ectomycorrhizal spores (Maser *et al.* 1978).

However, movement of biota from residual patches into burned areas is not always beneficial. These patches can also house pathogens. For example, dwarf mistletoe (*Arceuthobium* spp.), a parasitic plant that occurs throughout the North American boreal forest and infests primarily pines but also spruces and balsam fir, can survive in these patches and spread to the regenerating forest (Alexander and Hawksworth 1976, Hawksworth and Wiens 1996).

Supplementing regeneration of trees

As discussed previously, much of the post-fire regeneration that occurs in burned areas originates *in situ* based on regeneration from propagules that escaped or survived the effects of the fire (i.e., that endured or evaded the fire). However, there are also invaders that regenerate through wind-dispersed seeds, spores, or fragments that migrate into and regenerate in the burned areas. Residual patches within the wildfire footprint appear to provide a ready source of these propagules.

Some authors (e.g., Heinselman 1981) have mentioned the existence of these residual patches and their role in providing remnants of the pre-fire ecosystem that become a possible source of renewal of the subsequent system. Foster (1985) stated that the rate of post-fire tree regeneration depended on the size and pattern of the burn, since seedling establishment depended somewhat on propagules from residual patches within the wildfire footprint. Although we found only one study (Galipeau *et al.* 1997) that specifically evaluated the contribution of residual patches to regenerating the burned area, we can use what is known about autecology and post-fire succession, both topics that have been extensively discussed in the scientific literature, to piece together how these residual patches may contribute to establishing the new forest.

Many boreal tree species are not serotinous, do not maintain viable persistent soil seed banks in the long term, and do not reproduce vegetatively. Instead, most tree species, both coniferous and deciduous, rely on wind-dispersed seeds (Johnson 1992), even though most of these seeds do not disperse over long distances. Residual patches within the wildfire footprint could therefore be a vital source of propagules for their regeneration. For example, white spruce (*Picea glauca*), balsam fir (*Abies balsamifera*), larch (*Larix* spp.), and northern white cedar (*Thuja occidentalis*) would need to rely on seed from residual patches or the edges of the burn to regenerate (Dix and Swan 1971, Viereck *et al.* 1979). White spruce regeneration would depend on its ability to seed in from residual

ECOLOGICAL ROLES OF WILDFIRE RESIDUALS

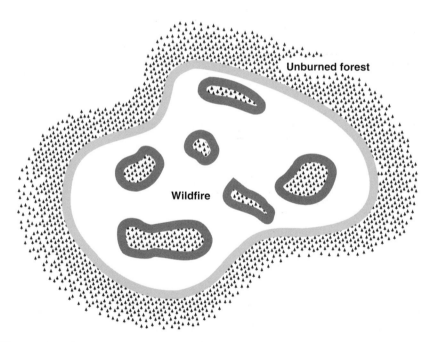

Figure 4.14 Residual patches within wildfires provide a source of seed to support the establishment of the next forest. When these unburned areas are abundant and well interspersed, they effectively increase the seed-dispersal distance within the wildfire footprint by providing internal seed sources. In the scenario illustrated, if the residual patches did not exist, only the area shaded in light grey could regenerate from wind-dispersed seed. On account of the residual patches, this area may be enlarged (dark grey areas).

patches or surrounding unburned forest, as its seed dispersal range is less than 100 m (Galipeau *et al.* 1997, Greene and Johnson 2000). Furthermore, its seeds are only viable for 1 year (Zasada and Gregory 1969). This indicates that the spatial distribution of residual patches within a wildfire footprint is likely to be an important determinant for the regeneration success of white spruce and possibly other species (Figure 4.14). Even for birch species, which are prolific sprouters, seeding in from adjacent unburned areas can be important for post-fire regeneration (Ahlgren 1959). Since birch seed dispersal distance ranges from 100 to 200 m, residual patches within the wildfire footprint may be an important source of propagules. Since black spruce seed dispersal is limited to distances of one or two times the height of the tree (LeBarron 1939, Heinselman 1957), short-distance seed dispersal from residual patches is likely to be the primary source of regeneration when black spruce cones in the burned area are completely destroyed. As Johnson (1992) noted, the role of residual patches as a proximal source of tree seeds for post-fire recovery may be considerably more important in larger burns, where the seed from *outside* the burned area would not reach most internal parts.

As detailed in Chapter 2, which discusses the formation of residual vegetation, residual patches may be representative of the pre-fire ecosystem, or they may form a subset representing some component of that system (e.g., wetland species that escape because their habitat is too wet to burn). Whether they represent the whole ecosystem or only a part is likely to greatly influence the species composition of the post-fire regeneration.

Providing connections to areas outside the wildfire footprint

In addition to acting as a direct source of biota for recolonizing the burn, and reducing the distance over which propagules must flow (i.e., by serving as internal sources), it is possible that residual patches may form a network through which biota flow into the burn from *outside* the wildfire footprint. If the residual patches are arranged in a suitable manner, they could offer functional connectivity to species that could not, or would prefer not to, migrate great distances across exposed burned areas. We have observed this kind of network distribution after many boreal wildfires in Canada (Figure 4.15).

Unfortunately, the functional effectiveness of these networks has not been documented, except for a mention by Gandhi (1999), who indicated that the presence of patchily distributed residual patches of variable size may both provide habitat and increase connectivity between habitats for endemic beetle species.

Overall, the role of residual patches as a source of regeneration material during the recovery process has not been well studied, although the available evidence suggests they are important while burned areas recover their original plant,

ECOLOGICAL ROLES OF WILDFIRE RESIDUALS

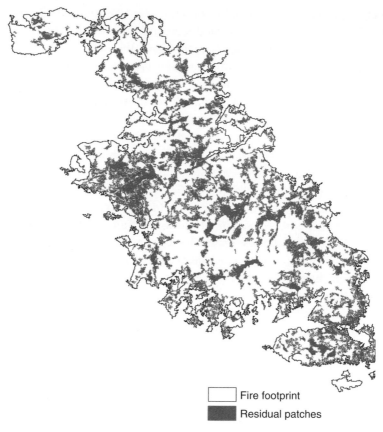

Fire footprint

Residual patches

Figure 4.15   An example of a boreal wildfire footprint showing the internal network of unburned areas. This 3750-ha fire contained almost 30% of its area in residual patches (Perera *et al.* 2009). These residual patches may enhance the connectivity within the wildfire footprint, making the center of the burned area more accessible for biotic flows from outside the wildfire footprint.

animal, and microorganism species assemblages. Also, the flow of adult and juvenile animals from residual patches to burned areas is not well understood. However, there is much indirect evidence that this flow occurs for lichens, ectomycorrhizae, arthropods, amphibians, and small mammals. Equally unknown are the effects of the spatial characteristics of residual patches (e.g., size, shape, and distribution) on their role, and changes over time after the fire add to the complexity. We also do not know what role the residual patches play as stepping stones – as nodes in a network – that facilitate the flow of biota into the burned areas, or the influence of the sizes, spatial configurations, and proximity of these patches to the edges of the fire.

ECOLOGICAL ROLES OF WILDFIRE RESIDUALS

Creating heterogeneity

Periodic wildfires create a shifting spatial mosaic of forest cover types and age classes, and this is an inherent characteristic of boreal forest landscapes. This diversity in land cover patterns is important for providing habitat for many animal species in the boreal landscape, and even for the livelihood of northern rural communities who depend on non-timber products from the forest (Nelson *et al.* 2008). What is probably less known is the heterogeneity created by the occurrence and spatial distribution of residual patches within wildfire footprints, which manifest at a smaller scale than the heterogeneity created by consecutive fires. As Gandhi (1999) demonstrated with beetles, this heterogeneity of residual patches translates into increased species diversity within a wildfire footprint. As we noted earlier, large fires typically produce many residual patches, and the resulting spatial patterns last for many decades (e.g., Arsenault 2001), even after the burned areas have recovered their forest cover. There is some evidence that the habitat provided by residual patches may also last many decades. For example, Seip and Parker (1997) found that even 30 – 40 years after a fire, residual patches contained the same bird species found in the unburned forest outside the wildfire footprint.

However, the contribution of this spatial and temporal diversity in habitat to ecological processes has not been fully explored and, as Nelson *et al.* (2008) observed, the socio-economic consequences for rural communities in boreal landscapes are equally unknown. In summary, the interspersion of unburned and burned areas adds another dimension to the landscape heterogeneity created by recurring boreal wildfires. They result in a spatial mosaic of ages and species compositions within the wildfire footprints, at scales ranging from patches of trees to small islands of unburned sphagnum moss. This range of microsites creates a heterogeneous landscape that is likely to result in high diversity of species and ecosystems. In addition, less-mobile animals that survived within residual patches may contribute to long-term genetic diversity, especially if the residual patches are large (and formed due to specific geo-environmental factors that protected them) and if the fire-return interval is long.

Ecological significance

In summary, the contributions of patches of residual vegetation may span multiple spatial and temporal scales over long periods. Their immediate role after a fire is to provide shelter for mobile animal species that flee the fire. This role begins at the time of the fire and may last for only a brief period thereafter. The scope of this effect may remain mostly local, but if the extent of residual patches is large and if the patches are well distributed with respect to the burned areas, the sheltering role can be significant. Over time, residual patches assume a greater role in supplementing the recovery of the burned area from the time of the fire until the burned area matures and attains the habitat characteristics required by a range of species.

The importance of this provision of a refuge would depend on the species in question, and on the size and species composition of the residual patches. However, it is clear that residual patches become a source for dispersal that aids the recovery process, as a proximal source of plant propagules and animal emigrants to the wildfire footprint. Less obvious is the value of a network of well-distributed residual patches within a wildfire footprint. This network increases the number of pathways by which biotic material can flow inward from outside the wildfire footprint. Thus, residual patches effectively shrink the size of the burned area because they act as stepping stones that reduce the dispersal distance from a source to a destination.

We contend that the ecological role of residual patches continues long after the burned system has recovered. The legacies of the residual patches contribute to the spatial mosaic of patches with different ages and compositions, and thereby contribute to heterogeneity at a broader spatial scale. Beyond their value for habitat supply and species diversity, they may have evolutionary consequences. For example, if the residual patches repeatedly escape fire due to strong spatial biases that cause them to form, they will remain part of the landscape for long periods, and may sustain unique species assemblages that differ from those in the periodically burned parts of the landscape.

We do not intend to imply that such roles are restricted to large unburned areas of different forest types. In fact, they are relevant to residual patches of all vegetation types, and at all sizes ranging from smaller patches of shrubs or sphagnum moss that escape the fire locally to large areas of forest. There is evidence to suggest that depending on the species and ecological process in question, residual patches can play important roles at all spatial scales. The degree to which residual patches influence the processes described above will depend on their composition, individual size, total extent, and spatial distribution within the wildfire footprint.

Wildfire residuals and the carbon cycle

The effect of wildfires on carbon cycles is drawing considerable attention from both scientists and policymakers, and nowhere is this concern more prominent than in boreal forest landscapes. The focus on the pan-boreal forest biome is not surprising given that it may store up to one-third of the global carbon pool (e.g.,

Smith *et al*. 1993, Pan *et al*. 2011) and that intense wildfires occur frequently in this biome. Many studies have been conducted on this topic, in particular to estimate the carbon balance of the boreal biome, using both empirical and simulation modeling techniques (Kasischke and Stocks 2000). Though the results vary considerably, and though debates continue over the study methods and assumptions, most researchers appear to agree that there will be a net carbon loss from the boreal forest, especially if global warming increases the frequency of wildfires (Harden *et al*. 2000, Bond-Lamberty *et al*. 2007). This release of carbon from boreal wildfires at a massive scale is also expected to create positive feedback that intensifies climate change, since the major forms of carbon emission are the greenhouse gases carbon dioxide (mostly) and methane (Kasischke *et al*. 1995).

The role of residual vegetation in boreal wildfires is an important consideration in this context from two viewpoints. First, there is a methodological point that relates to how carbon cycle estimates are computed; secondly, there is the functional role of residuals in storing carbon in the post-fire ecosystem. Global and regional carbon budgets are typically developed using models and extrapolations from empirical information (Kasischke and Stocks 2000). This is also true for estimates of carbon emission from boreal wildfires and, as Goldammer and Furyaev (1996) pointed out for boreal Eurasia, such estimates are based on official fire statistics and fuel consumption data with questionable reliability. This concern was echoed more recently for boreal North America by Kasischke and Hoy (2012), who noted that the inconsistencies in estimates of carbon emission from wildfires were as high as 20%.

These inconsistencies stem from how wildfire footprints are mapped (as discussed in Chapter 3) as well as what is included in the estimates of burned area. The information required to improve the reliability of such computations includes accurate estimates of the unburned area within the mapped fire perimeters. For example, the "large fire database" used in Canada for such purposes has no consistent estimate of the unburned area within fire perimeters, since the mapping methods vary from fire to fire and among the jurisdictions in charge of mapping and processing fire maps (Amiro *et al*. 2001). Assumptions of low percentages of unburned area within large fires (e.g., 3–5% by Amiro *et al*. 2001) may be a gross underestimation of the area of the residual patches remaining after boreal wildfires. For example, Kasischke and Hoy (2012) estimated that 20% of the area affected by wildfires was unburned based on data from 75 fire events ranging from 500 to more than 215 000 ha in Alaska (USA). As we pointed out in Chapter 3, boreal wildfires commonly produce residual patches that occupy up to a quarter of the land area within a fire perimeter, and that percentage does not seem to decrease as the size of the fire increases.

Although the percentage of unburned area may vary among fires, the important point is that it will not be an insignificant value, and underestimation of the

ECOLOGICAL ROLES OF WILDFIRE RESIDUALS

actual unburned area would lead to considerable exaggeration of the magnitude of carbon emission. For instance, the assumption that only 3% of the wildfire footprint was occupied by unburned vegetation and that 80 million ha were affected by wildfire in Canada from 1959 to 1999 (Amiro *et al.* 2001) produces an estimated 2.4 million ha of unburned forest; if 20% of the area is not burned, this produces 16 million ha of unburned forest, and the difference between the two estimates represents a vast underestimation of the importance of residuals in storing carbon, and an equally vast corresponding overestimation of carbon emission.

Of course, the actual error in such estimates depends on the characteristics of the unburned area and its carbon stock, both above and below the ground. Even so, it is surprising that even some recent assessments of the fate of forest carbon stocks subject to wildfires have not considered residuals within the burns as a significant source of variability (e.g., Nave *et al.* 2011). We contend that the area of the residual patches must not be ignored when computing the carbon emission from boreal wildfires, since this would lead to significant overestimation of carbon release from this highly fire-prone region. As French *et al.* (2008) demonstrated, burn severity indices derived from satellite imagery could provide reliable estimates of burn heterogeneity and of the unburned areas within fires, and could thus lead to better carbon emission estimates. Furthermore, in our effort to ascertain the carbon stocks remaining in residual patches after fires from the scientific literature, we discovered that simple descriptions of carbon stocks in boreal forest landscapes in terms of common cover types (e.g., black spruce–sphagnum moss, jack pine–lichen woodland) are not readily available. After perusing many publications on carbon content and dynamics in boreal landscapes, we could not find any generalized carbon profiles associated with boreal forest cover types. Without such baseline information, it is difficult to estimate the residual carbon stocks in wildfire footprints with any degree of generalization.

The role of snag residuals and downed trees that do not completely turn into ash within the burned area also cannot be neglected (Figure 4.16). As described in Chapter 3, this material constitutes a large volume of wood even after the most intense of boreal wildfires. For example, Balshi *et al.* (2009) estimated that the aboveground fraction of carbon consumed in boreal wildfires ranged from 13 to 26%; this indicates that 74% or more of the arboreal carbon remains intact. The aerial biomass lost (mostly fine branches and needles) from snag residuals during boreal wildfires is negligible, and the biomass not consumed (mostly stems and coarse branches) is substantial. Consequently, these snag residuals trap a large mass of carbon even after the fire, with values ranging from 75 to 80% of the aerial biomass in unburned stands (Ohmann and Grigal 1979, Slaughter *et al.* 1998).

Specific estimates of these carbon stores in wood that survives the fire (snag residuals and downed wood) vary based on the fire severity and many characteristics of the pre-burn forest community, such as tree age, species composition,

Figure 4.16   Much of the carbon contained in pre-fire vegetation remains on site following even intense boreal wildfires. This photograph was taken 1 year after a boreal wildfire.

and site factors (Slaughter *et al.* 1998, Bond-Lamberty *et al.* 2003, Boulanger and Sirois 2006, Boulanger *et al.* 2011, Brown and Johnstone 2011). In contrast to the sudden pulse of carbon released during a wildfire, carbon stored in snag residuals and downed wood is released slowly through decomposition, often over many decades. The release rates of this stored carbon remain low, but may change over time in response to seasonal environmental changes, variations in site factors such as soil moisture, and differences in the sizes and nutrient contents of stems (Bond-Lamberty *et al.* 2003, Boulanger and Sirois 2006). The reservoir of carbon in boreal wildfire residuals was not considered in global carbon budget estimates until recently, but recognition of its importance has made it an increasingly popular topic of study (Harden *et al.* 2000, Kasischke 2000). Furthermore, when standing burned wood falls, it creates a large mass of downed wood, which will gradually be buried and become part of the soil, forming deep soil storage of carbon. Some estimates suggest that this source of carbon contributes as much as 60% of the deep soil carbon reserves at post-fire boreal sites (Manies *et al.* 2005).

Wildfire residuals and nutrient and hydrological cycles

Wildfires release minerals from the soil, the forest floor, and the aboveground vegetation due to the direct effects of heating and combustion, as described in the

ECOLOGICAL ROLES OF WILDFIRE RESIDUALS

comprehensive reviews by DeBano *et al.* (1998) and Certini (2005). The type and the amount of nutrients released in this way depend on many variables, including fire intensity, forest composition, and season of the burn (Chandler *et al.* 1983). One prominent change is the formation of a water-repellent layer at the soil surface, especially after fires of medium to high intensity (DeBano *et al.* 1998, Moore and Richardson 2012). This layer increases surface runoff of water – a "tin roof" effect – for some time after the fire. This is a prevalent post-fire condition in boreal forest soils, as is the formation of a hardened crust on the mineral soil as a result of thermal stress caused by the fire, the oxidation of organic matter, and subsequent freezing and thawing (Kelsall *et al.* 1977).

Due to the losses of cover by the forest canopy and of forest-floor litter, the structure capable of intercepting precipitation in burned areas is greatly reduced after a fire. The combination of an increased proportion of the precipitation reaching the soil and the formation of a water-repellent soil layer promotes surface runoff that will carry away fine sediments. In combination with the post-fire mobilization of minerals that were formerly held in organic matter, this produces a flush of nutrients into wetlands, streams, and lakes. It is well known that wildfires increase streamflow, and decrease the physical quality of the water, through increased turbidity and sediments, and the chemical quality of the water, through increased levels of dissolved nutrients (DeBano *et al.* 1998, Moore and Richardson 2012).

Boreal wildfires also influence the amount of sediment found in brooks and streams (Beaty 1994) and thereby increase silt loads and dissolved chemical concentrations (Schindler *et al.* 1980, Bayley *et al.* 1992). This phenomenon may have consequences for plants and animals in aquatic ecosystems. For example, St-Onge and Magnan (2000) found that boreal wildfires decreased the abundance of small yellow perch (*Perca flavescens*) and white suckers (*Catastomus commersoni*) in lakes in Québec (Canada) through increased sedimentation. Algal biomass also increases in lakes in proportion to the catchment area disturbed by wildfires as a consequence of increased nutrient loading (Planas *et al.* 2000). Such effects accumulate and affect ecological processes at a watershed scale, as detailed by Pinel-Alloul *et al.* (2002) in a review of boreal watersheds affected by fire disturbance.

The vegetation that occurs at the land–water interface around streams and lakes, which is generally referred to as the riparian zone, is known to influence many ecological processes at local and landscape levels. As we described earlier, many of these riparian zones are left unburned during boreal wildfires. These zones act as a buffer against nutrient fluxes and sediment loading in the water bodies they border. However, the details of how these processes are mitigated by riparian zones are not well understood, not only for boreal forests but also across all other major vegetation zones (Pettit and Naiman 2007). The lack of information may result from the difficulties in the experimental design (i.e., advanced planning and replication) required to study post-fire effects in watershed studies as well as the difficulty of providing experimental evidence of direct contributions by the *presence* of residual unburned riparian areas. Consequently, the effects of forest fire on boreal riparian zones and water bodies are typically studied opportunistically after wildfires (e.g., Cott *et al.* 2010).

Although nutrient cycling soon after a wildfire may be mostly an abiotic process, wood-boring beetles and woodpeckers that occupy snag residuals can accelerate this process. Production and scattering of frass (wood chips and fecal matter) by wood-boring beetles that inhabit the snag residuals enhance the return of nutrients to the forest floor during the first several years after a fire, and even assist soil microbial activity. Furthermore, the deep feeding galleries produced by beetle larvae in snags appear to encourage fungal growth and subsequent decay of the snag residuals, thereby accelerating return of the stored nutrients.

ECOLOGICAL ROLES OF WILDFIRE RESIDUALS

The evidence currently available comes from studies on the negative effects of the *absence* of residual vegetation. For example, the removal of vegetation around water bodies has been found to decrease spawning habitat quality by increasing the amount of fine suspended sediment that may be carried to a shoreline by water movement and be deposited in spawning grounds. This has led to the hypothesis that buffers around lakes and streams may reduce such effects (St-Onge and Magnan 2000). Similarly, Planas *et al.* (2000) stated that the lack of riparian vegetation in some burnt watersheds might explain why the responses of biomass were greater and the algal composition was more complex in lakes surrounded by a burned area than in lakes surrounded by a harvested area. In such cases, algal biomass may not reach a steady state even 3 years after the disturbance. France (1997) also argued that the removal of riparian vegetation by wildfires in boreal lakes of the Canadian Shield would raise water temperature and lower the thermocline due to exposure of the water bodies to increased sunlight and wind effects. This would affect fish that require cold water such as lake trout (*Salvalinus namaycush*) (France 1997) and those that are sensitive to both temperature and light conditions such as walleye (*Sander vitreus*) (Lester *et al.* 2004). The increased exposure to solar radiation can also increase the rate of snowmelt, leading to an increased input of cold meltwater (Kelsall *et al.* 1977).

In summary, there are many plausible ways that residual vegetation can mitigate the effects of increased runoff, sedimentation, and nutrient loading of aquatic systems, both within the areas burned by boreal wildfires and downstream, at a watershed scale (Figure 4.17). Given the paucity of direct empirical evidence, however, such effects remain untested hypotheses, and their probabilities and variability in relation to the species composition of the original forest and to geo-environmental factors remain uncertain.

Summary

We have attempted to encapsulate the state of knowledge of the ecological roles of residual vegetation after a boreal wildfire based on a review of the scientific literature. Table 4.3 provides a brief, synopsis of what is known and what remains uncertain. Our main finding is that the current knowledge appears to be mostly based on studies that focused on a given species or species group (e.g., sawyer beetles or the black-backed woodpecker) or a specific ecosystem process (e.g., the decay of downed wood or tree seed dispersal), and these details comprise what is well known. Overall, the ecological role of snag residuals in terms of their use by saproxylic beetles and woodpeckers and their role in regeneration and carbon storage seem to be well known through much dedicated research by entomologists, ornithologists, and plant ecologists. In contrast, not much is known about residuals in terms of their holistic role, for example, in the recovery of the burned ecosystem or in mitigating the deleterious effects of wildfire. Studies on the role of residual patches appear to be particularly lacking, which is surprising given both the wealth of studies on fire ecology and early acknowledgment of the omnipresence of residual patches (e.g., Rowe and Scotter 1973) and their potential ecological roles (Johnson 1992).

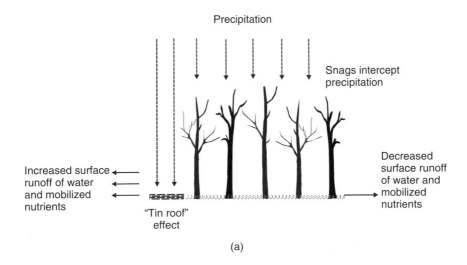

Precipitation

Snags intercept
precipitation

Increased surface
runoff of water
and mobilized
nutrients

Decreased
surface runoff
of water and
mobilized
nutrients

"Tin roof"
effect

(a)

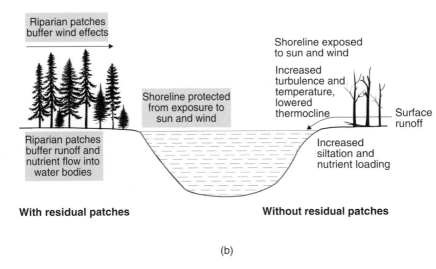

Riparian patches
buffer wind effects

Shoreline exposed
to sun and wind

Increased
turbulence and
temperature,
lowered
thermocline

Shoreline protected
from exposure to
sun and wind

Surface
runoff

Riparian patches
buffer runoff and
nutrient flow into
water bodies

Increased
siltation and
nutrient loading

With residual patches

Without residual patches

(b)

ECOLOGICAL ROLES OF WILDFIRE RESIDUALS

Figure 4.17   Residual vegetation after a boreal forest wildfire may mitigate the effects of the disturbance on bodies of water within an affected watershed: (a) snag residuals can intercept precipitation, thereby reducing surface runoff, and (b) residual riparian patches can buffer streams and lakes against the transport of sediments and nutrients by overland flow. Also, the vegetation protects the water from sun, minimizing temperature fluctuations in the lower layers, and blocks the wind, reducing turbulence and mixing.

Table 4.3 Summary of what is known and what is uncertain with respect to the ecological roles of boreal wildfire residuals.

Residual type and ecological role	What is well known	Major gaps in knowledge and uncertainties
Snags		
Post-fire animal species dynamics	Saproxylic beetles and woodpeckers, invasion processes, and life cycles at a local scale	Other species that occupy snag residuals early in the recovery process Regional-level roles of snag residuals initially as a target for occupation and later as a source for dispersal of saproxylic beetles and woodpeckers Woodpecker population dynamics at larger temporal and spatial scales Role of successive fires and the unburned matrix in maintaining populations of fire-dependent species Effects of the regional fire cycle on population levels; secondary cavity nesters
Supporting the recovery of the burned system	Supply of propagules through serotinous cones of some conifers, and vegetative propagules (e.g., roots and stem sprouts) of deciduous trees and shrubs	
Carbon cycle	Capacity to form an aboveground reservoir of carbon	Contribution to the belowground reservoir of carbon Magnitude of the aboveground reservoir and its actual carbon content
Hydrology and nutrient cycle	Nutrient pool in the aboveground stems and release through the decomposition of downed wood	Interception of precipitation by snag residuals and influence on surface and subsurface runoff Role of beetles in accelerating nutrient availability for regeneration

(*continued overleaf*)

Table 4.3 (*continued*)

Residual type and ecological role	What is well known	Major gaps in knowledge and uncertainties
Residual patches		
Post-fire species dynamics		Shelter for mobile animals during fires
		Longer-term refuges (lifeboating) for species that use the mosaic of burned and unburned patches
		Dispersal source for late-successional species
		Contribution to long-term germplasm conservation and ecosystem heterogeneity at regional scales
Supporting recovery of the burned system		Formation of a network for animal movement from outside the wildfire footprint
		Dispersal of propagules into the burned areas by shrubs, mosses, lichens, and ectomycorrhizae
		Effect of size, and extent and spatial distribution of the patches on post-fire species dynamics and recovery of the burned system
Carbon cycle		Extent of carbon stored in residual patches in relation to regional and global carbon budgets
Hydrology and nutrient cycles		Riparian buffer effects at fire-footprint and watershed scales:
		mitigating post-fire runoff, sedimentation, and nutrient loading of streams and lakes;
		sheltering lakes from heating and exposure to wind;
		minimizing negative consequences on fish and other aquatic organisms

ECOLOGICAL ROLES OF WILDFIRE RESIDUALS

Furthermore, although a synoptic understanding of the ecological processes related to wildfire residuals can be logically constructed to some extent from the findings in studies of individual components, no formal attempts have been made to develop such a synthesis. Consequently, any such constructs remain informal hypotheses and often take the form of assumptions, such as the belief that residual patches are necessary to buffer bodies of water against sedimentation after wildfires or that a continuous, spatially and temporally well-distributed supply of snags is necessary to conserve the regional diversity of saproxylic beetles. In particular, there are considerable gaps in the knowledge of the ecology of residuals at broader scales, for example, of their role beyond the wildfire footprint. The list of what is not well known in Table 4.3 is by no means exhaustive; it is only what we considered to be the most important examples from a broader point of view. Those who study individual species and specific processes can certainly add to this list, both in terms of what is well known in their field of research and in terms of uncertainties. Our purpose in presenting this table is to stimulate discussion, and even debate, that will guide future studies of residuals.

Interest in understanding the ecological processes associated with wildfire residuals is more than merely academic. As we discuss in detail in Chapter 5, this information is used to guide forest management from the perspective of conservation biology, which sometimes conflicts with economic and social interests. Therefore, it is logical to examine and understand what is well known about the ecology of wildfire residuals, as well as the major gaps and uncertainties in knowledge.

Some questions about boreal wildfire residuals for researchers:

1. How does the spatio-temporal distribution of wildfires (i.e., the supply of snag residuals) affect the regional population dynamics of saproxylic beetles and woodpeckers?
2. What plant and animal species "lifeboat" in residual patches and how do the patch sizes and distribution of unburned areas affect this process?
3. How does the extent of residual patches and their spatial interspersion with burned patches influence the regeneration of plant species and the recovery of animal populations within areas burned by wildfires?
4. How do the extent and spatial patterns of residual patches form a network to enable the flow of biota into the burned area from outside the area of the wildfire?
5. What is the contribution to long-term germplasm conservation of residual patches that repeatedly escape fire?
6. How do we quantify the amount of carbon stored in residual patches that escape wildfires and incorporate these estimates into estimates of the regional carbon flux?
7. How important is the presence of riparian residual patches in controlling nutrient loading and sedimentation of lakes in the watershed downstream of a fire?

References

Ahlgren, C.E. (1959) Some effects of fire on forest reproduction in northeastern Minnesota. Journal of Forestry 57, 194-200.

Ahlgren, C.E. (1974) Effect of fires on temperate forests: north central United States. In: Kozlowski, T.T. (ed.) Fires and Ecosystems. Academic Press, New York, NY. pp. 195–219.

Ahnlund, H. and Lindhe, A. (1992) Endangered wood-living insects in coniferous forests – some thoughts from studies of forest-fire sites, our crops and clearing in the province of Sörmland, Sweden. Enromologisk Tidskrift 113, 13–23 (in Swedish).

Alexander, M.E. and Hawksworth, F.G. (1976) Fire and dwarf mistletoes in North American coniferous forests. Journal of Forestry 74, 446-449.

Amiro, B.D., Todd, J.B., Wotton, B.M. et al. (2001) Direct carbon emissions from Canadian forest fires, 1959-1999. Canadian Journal of Forest Research 31, 512–525.

Angelstâm, P. and Mikusinski, G. (1994) Woodpecker assemblages in natural and managed boreal and hemiboreal forest – a review. Annales Zoologici Fennici 31, 157–172.

Arseneault, D. (2001) Impact of fire behavior on postfire forest development in a homogeneous boreal landscape. Canadian Journal of Forest Research 31, 1367–1374.

Auclair, A.N.D. (1985) Postfire regeneration of plant and soil organic pools in a *Picea mariana–Cladonia stellaris* ecosystem. Canadian Journal of Forest Research 15, 279–291.

Axtell, H.H. (1957) The three-toed woodpecker invasion. Audobon Outlook 6, 13–14.

Baldwin, P.H. (1968) Predator-prey relationships of birds and spruce beetles. In Proceedings of the North Central Branch, Entomological Society of America, Denver, CO, Vol. 23. pp. 90–99.

Balshi, M.S., McGuire, A.D., Duffy, P., Flannigan, M. Kicklighter, D.W., and Melillo, J. (2009) Vulnerability of carbon storage in North American boreal forests to wildfires during the 21st century. Global Change Biology 15, 1491–1510.

Bayley, S.E., Schindler, D.W., Parker, B.R., Stainton, M.P., and Beaty, K.G. (1992) Effects of forest fire and drought on acidity of a base-poor boreal forest stream: Similarities between climatic warming and acidic precipitation. Biogeochemistry 17, 191–204.

Beaty, K.G. (1994) Sediment transport in a small stream following two successive forest fires. Canadian Journal of Fisheries and Aquatic Sciences 51, 2723–2733.

Beaufait, W. (1960) Some effects of high temperatures on cones and seeds of jack pine. Forest Science 6(3), 194–199.

Blackford, J.L. (1955) Woodpecker concentration in burned forest. The Condor 57, 28–30.

Bock, C.E. and Lynch, J.F. (1970) Breeding bird populations of burned and unburned conifer forest in the Sierra Nevada. The Condor 72, 182–189.

Bonan, G.B. and Shugart, H.H. (1989) Environmental factors and ecological processes in boreal forests. Annual Review of Ecology and Systematics 20, 1–28.

Bond-Lamberty, B.P., Peckham, S.D., Ahl, D.E., and Gower, S.T. (2007) Fire as the dominant driver of central Canadian boreal forest carbon balance. Nature 450(7166), 89–89.

Bond-Lamberty, B., Wang, C., and Gower, S.T. (2003) Annual carbon flux from woody debris for a boreal black spruce fire chronosequence. Journal of Geophysical Research-Atmospheres 108(D23), 8220–8230.

Bonnot, T.W., Rumble, M.A., and Millspaugh, J.J. (2008) Nest success of black-backed wood-peckers in forests with mountain pine beetle outbreaks in the Black Hills, South Dakota. The Condor 110, 450–457.

Boulanger, Y. and Sirois, L. (2006) Postfire dynamics of black spruce coarse woody debris in northern boreal forest of Quebec. Canadian Journal of Forest Research 36, 1770–1780.

Boulanger, Y. and Sirois, L. (2007) Postfire succession of saproxylic arthropods, with emphasis on Coleoptera, in the north boreal forest of Quebec. Environmental Entomology 36, 128–141.

Boulanger, Y., Sirois, L., and Hebert, C. (2011) Fire severity as a determinant factor of the decomposition rate of fire-killed black spruce in the northern boreal forest. Canadian Journal of Forest Research 41, 370–379.

Bradbury, S.M. (2006) Response of the post-fire bryophyte community to salvage logging in boreal mixedwood forests of northeastern Alberta, Canada. Forest Ecology and Management 234, 313–322.

Brown, C.D. and Johnstone, J.F. (2011) How does increased fire frequency affect carbon loss from fire? A case study in the northern boreal forest. International Journal of Wildland Fire 20, 829–837.

Buddle, C.M., Spence, J.R., and Langor, D.W. (2000) Succession of boreal forest spider assemblages following wildfire and harvesting. Ecography 23, 424–436.

Buech, R.R., Siderits, K., Radtke, R.E., Sheldon, H.L., and Elsing, D. (1977) Small mammal populations after a wildfire in northeast Minnesota. USDA Forest Service, North Central Research Station, St. Paul, MN. Research Paper NC-151.

Cayford, J.H. and McRae, D.J. (1983) The ecological role of fire in jack pine forests. In Wein, R.W. and MacLean, D.A. (eds) The Role of Fire in Northern Circumpolar Ecosystems. John Wiley & Sons, Ltd, New York, NY. pp. 183–199.

Certini, G. (2005) Effects of fire on properties of forest soils: a review. Oecologia 143, 1–10.

Chandler, C., Cheney, P., Thomas, P., Trabaud, L., and Williams, D. (1983) Fire in Forestry: Volume I. Forest Fire Behaviour and Effects. John Wiley & Sons, Ltd, New York, NY.

Cobb, T.P., Hannam, K.D., Kishchuk, B.E., Langor, D.W., Quideau, S.A., and Spence, J.R. (2010) Wood-feeding beetles and soil nutrient cycling in burned forests: Implications of post-fire salvage logging. Agricultural and Forest Entomology 12, 9–18.

Cobb, T.P., Langor, D.W., and Spence, J.R. (2007) Biodiversity and multiple disturbances: boreal forest ground beetle (Coleoptera: Carabidae) responses to wildfire, harvesting, and herbicide. Canadian Journal of Forest Research 37, 1310–1323.

Cott, P.A., Zaidlik, B.A., Bourassa, K., Lange, M., and Gordon, A.M. (2010) Effects of forest fire on young-of-the-year northern pike, *Esox lucius*, in the Northwest Territories. Canadian Field-Naturalist 124, 104–112.

Dahlberg, A. (2002) Effects of fire on ectomycorrhizal fungi in Fennoscandian boreal forests. Silva Fennica 36, 69–80.

Dahlberg, A., Schimmel, J., Taylor, A.F.S., and Johannesson, H. (2001) Post-fire legacy of ectomycorrhizal fungal communities in the Swedish boreal forest in relation to fire severity and logging intensity. Biological Conservation 100, 151–161.

DeBano, L.F., Neary, D.G., and Ffolliott, P.F. (1998) Fire's Effects on Ecosystems. John Wiley & Sons, Ltd, New York, NY.

de Chantal, M. and Granström, A. (2007) Aggregations of dead wood after wildfire act as browsing refugia for seedlings of *Populus tremula* and *Salix caprea*. Forest Ecology and Management 250, 3–8.

DeLong, S.C. and Kessler, W.B. (2000) Ecological characteristics of mature forest remnants left by wildfire. Forest Ecology and Management 131, 93–106.

Dettki, H., Klintberg, P., and Esseen, P.A. (2000) Are epiphytic lichens in young forests limited by local dispersal? Ecoscience 7, 317–325.

Dix, R.L. and Swan, J.M.A. (1971) The roles of disturbance and succession in upland forest at Candle Lake, Saskatchewan. Canadian Journal of Botany 49, 657–676.

Edmonds, R.L. and Eglitis, A. (1989) The role of the Douglas-fir beetle and wood borers in the decomposition of and nutrient release from Douglas-fir logs. Canadian Journal of Forest Research 19, 853–859.

Ellison, L.N. (1975) Density of Alaskan spruce grouse before and after fire. Journal of Wildlife Management 39, 468–471.

Engelmark, O. (1993) Early post-fire tree regeneration in a *Picea–Vaccinium* forest in northern Sweden. Journal of Vegetation Science 4, 791–794.

Evans, W.G. (1964) Infra-red receptors in *Melanophila acuminata* De Geer. Nature 202(492), 211.

Evans, W.G. (1966) Perception of infrared radiation from forest fires by *Melanophila acuminata* De Geer (Buprestidae, Coleoptera). Ecology 47, 1061–1065.

Eyre, F.H. (1938) Can jack pine be regenerated without fire? Journal of Forestry 36, 1067–1072.

Eyre, F.H. and LeBarron, R.K. (1944) Management of jack pine stands in the Lake States. USDA Forest Service, Washington, DC. Technical Bulletin No. 863.

Fayt, P. (2003) Insect prey population changes in habitats with declining vs. stable three-toed woodpecker *Picoides tridactylus* populations. Ornis Fennica 80, 182–192.

Fayt, P., Machmer, M.M., and Steeger, C. (2005) Regulation of spruce bark beetles by woodpeckers – a literature review. Forest Ecology and Management 206, 1–14.

Fisher, J.T. and Wilkinson, L. (2005) The response of mammals to forest fire and timber harvest in the North American boreal forest. Mammal Review 35, 51–81.

Foster, D.R. (1985) Vegetation development following fire in *Picea mariana* (black spruce)–*Pleurozium* forests of south-eastern Labrador, Canada. Journal of Ecology 73, 517–534.

France, R. (1997) Land-water linkages: influences of riparian deforestation on lake thermocline depth and possible consequences for cold stenotherms. Canadian Journal of Fisheries and Aquatic Sciences 54, 1299–1305.

Franklin, J.F., Lindemeyer, D., MacMahon, J.A. *et al.* (2000) Threads of continuity. Conservation Biology in Practice 1, 8–16.

French, N.H.F., Kasischke, E.S., Hall, R.J. *et al.* (2008) Using Landsat data to assess fire and burn severity in the North American boreal forest region: an overview and summary of results. International Journal of Wildland Fire 17, 443–462.

Galipeau, C., Kneeshaw, D., and Bergeron, Y. (1997) Postfire regeneration of white spruce and balsam fir in relation to seedbed and distance from seed sources. Canadian Journal of Forest Research 27, 139–147.

Gandhi, K.J.K. (1999) The importance of fire-skips as biotic refugia and the influence of forest heterogeneity on epigaeic beetles in pyrogenic stands of the northern Rocky Mountains. MScThesis, University of Alberta, Edmonton, AB.

Gandhi, K.J.K., Spence, J.R., Langor, D.W. *et al.* (2001) Fire residuals as habitat reserves for epigaeic beetles (Coleoptera: Carabidae and Staphylinidae). Biological Conservation 102, 131–141.

Gardiner, L.M. (1957) Deterioration of fire-killed pine in Ontario and the causal wood-boring beetles. The Canadian Entomologist 89, 241–263.

Gasaway, W.C. and Dubois, S.D. (1985) Initial response of moose, *Alces alces*, to a wildfire in interior Alaska. Canadian Field-Naturalist 99, 135–140.

Gentry, D.J. and Vierling, K.T. (2008) Reuse of woodpecker cavities in the breeding and non-breeding seasons in old burn habitats in the Black Hills, South Dakota. American Midland Naturalist 160, 413–469.

Goldammer, J.G. and Furyaev, V.V. (1996) Fire in Ecosystems of Boreal Eurasia. Kluwer Academic Publishers, Dordrecht, Forestry Sciences Vol. 48.

Greene, D.F. and Johnson, E.A. (1999) Modelling recruitment of *Populus tremuloides*, *Pinus banksiana*, and *Picea mariana* following fire in the mixedwood boreal forest. Canadian Journal of Forest Research 29, 462–473.

Greene, D.F. and Johnson, E.A. (2000) Tree recruitment from burn edges. Canadian Journal of Forest Research 30, 1264–1274.

Greene, D.F., Noël, J., and Bergeron, Y. (2004) Recruitment of *Picea mariana*, *Pinus banksiana*, and *Populus tremuloides* across a burn severity gradient following wildfire in the southern boreal forest of Quebec. Canadian Journal of Forest Research 34, 1845–1857.

Grove, S.J. (2002) Saproxylic insect ecology and the sustainable management of forests. Annual Review of Ecology and Systematics 33, 1–23.

Haney, A., Apfelbaum, S., and Burris, J.M. (2008) Thirty years of post-fire succession in a southern boreal forest bird community. American Midland Naturalist 159, 421–433.

Hanks, L.M. (1999) Influence of the larval host plant on reproductive strategies of cerambycid beetles. Annual Review of Entomology 44, 483–505.

Hannah, K.C. and Hoyt, J.S. (2004) Northern hawk owls and recent burns: Does burn age matter? The Condor 106, 420–423.

Hannon, S.J. and Drapeau, P. (2005) Bird responses to burning and logging in the boreal forest of Canada. Studies in Avian Biology 30, 97–115.

Harden, J.W., Trumbore, S.E., Stocks, B.J. *et al.* (2000) The role of fire in the boreal carbon budget. Global Change Biology 6(Suppl. 1), 174–184.

Hawksworth, F.G. and Wiens, D. (1996) Dwarf mistletoes: Biology, pathology and systematics. USDA Forest Service, Washington, DC. Agriculture Handbook 709.

Hebblewhite, M., Munro, R.H., and Merrill, E.H. (2009) Trophic consequences of postfire logging in a wolf–ungulate system. Forest Ecology and Management 257, 1053–1062.

Heinselman, M.L. (1957) Silvical characteristics of black spruce. USDA Forest Service, Lake States Forest Experiment Station, St Paul, MN. Station Paper 45.

ECOLOGICAL ROLES OF WILDFIRE RESIDUALS

Heinselman, M.L. (1981) Fire and succession in the conifer forests of northern North America. In West, D.C., Shugart, H.H., and Botkin, D.B. (eds) Forest Succession: Concepts and Applications. Springer-Verlag, New York, NY. pp. 374–405.

Hobson, K.A. and Schieck, J. (1999) Changes in bird communities in boreal mixedwood forest: Harvest and wildfire effects over 30 years. Ecological Applications 9, 849–863.

Hooven, E.F. (1969) The influence of forest succession on populations of small animals in western Oregon. In Black, H.C. (ed.) Wildlife and Reforestation in the Pacific Northwest, Symposium Proceedings. Oregon State University, Corvallis, OR. pp. 30–34.

Hossack, B.R. and Corn, P.S. (2007) Responses of pond-breeding amphibians to wildfire: Short-term patterns in occupancy and colonization. Ecological Applications 17, 1403–1410.

Hossack, B.R., Corn, P.S., and Fagre, D.B. (2006) Divergent patterns of abundance and age-class structure of headwater stream tadpoles in burned and unburned watersheds. Canadian Journal of Zoology 84, 1482–1488.

Hoyt, J.S. and Hannon, S.J. (2002) Habitat associations of black-backed and three-toed woodpeckers in the boreal forest of Alberta. Canadian Journal of Forest Research 32, 1881–1888.

Hutto, R.L. (1995) Composition of bird communities following stand-replacement fires in Northern Rocky Mountain (U.S.A.) conifer forests. Conservation Biology 9, 1041–1058.

Ibarzabal, J. and Desmeules, P. (2006) Black-backed woodpecker (*Picoides arcticus*) detectability in unburned and recently burned mature conifer forests in north-eastern North America. Annales Zoologici Fennici 43, 228–234.

Imbeau, L., Savard, J.-P.L., and Gagnon, R. (1999). Comparing bird assemblages in successional black spruce stands originating from fire and logging. Canadian Journal of Zoology 77, 1850–1860.

Jayen, K., Leduc, A., and Bergeron, Y. (2006) Effect of fire severity on regeneration success in the boreal forest of northwest Québec, Canada. Ecoscience 13, 143–151.

Johansson, P. (2008) Consequences of disturbance on epiphytic lichens in boreal and near boreal forests. Biological Conservation 141, 1933–1944.

Johansson, T., Hjältén, J., Stenbacka, F., and Dynesius, M. (2010) Responses of eight boreal flat bug (Heteroptera: Aradidae) species to clear-cutting and forest fire. Journal of Insect Conservation 14, 3–9.

Johansson, P., Wetmore, C.M., Carlson, D.J., Reich, P.B., and Thor, G. (2006) Habitat preference, growth form, vegetative dispersal and population size of lichens along a wildfire severity gradient. The Bryologist 109, 527–540.

Johnson, E.A. (1992) Fire and Vegetation Dynamics: Studies from the North American Boreal Forest. Cambridge University Press, Cambridge, UK.

Johnstone, J.F., Chapin, F.S. III,, Foote, J., Kemmett, S., Price, K., and Viereck, L.A. (2004) Decadal observations of tree regeneration following fire in boreal forests. Canadian Journal of Forest Research 34, 267–273.

Jonsson, L., Dahlberg, A., Nilsson, M.C. Zackrisson, O., and Karen, O. (1999) Ectomycorrhizal fungal communities in late-successional Swedish boreal forests, and their composition following wildfire. Molecular Ecology 8, 205–215.

ECOLOGICAL ROLES OF WILDFIRE RESIDUALS

Kasischke, E.S. (2000) Processes influencing carbon cycling in the North American boreal forest. In Kasischke, E.S. and Stocks, B.J. (eds) Fire, Climate Change and Carbon Cycling in the Boreal Forest. Springer, New York, NY. Ecological Studies 138. pp. 103–110.

Kasischke, E.S., Christensen, N.L. Jr,, and Stocks, B.J. (1995) Fire, global warming, and the carbon balance of boreal forests. Ecological Applications 5, 437–451.

Kasischke, E.S. and Hoy, E.E. (2012) Controls on carbon consumption during Alaskan wildland fires. Global Change Biology 18, 685–699.

Kasischke, E.S. and Stocks, B.J. (2000) Fire, Climate Change, and Carbon Cycling in the Boreal Forest. Springer, New York, NY. Ecological Studies 138.

Kelsall, J.P, Telfer, E.S., and Wright, T.D. (1977) The effect of fire on the ecology of the boreal forest, with particular reference to the Canadian north: a review and selected bibliography. Canadian Wildlife Service, Ottawa, ON. Occasional Paper #32. CW 69-1/32.

Kotliar, N.B., Reynolds, E.W., and Deutschman, D.H. (2008) American three-toed woodpecker response to burn severity and prey availability at multiple spatial scales. Fire Ecology 4, 26–45.

Krefting, L.W. and Ahlgren, C.E. (1974) Small mammals and vegetation changes after fire in a mixed conifer-hardwood forest. Ecology 55, 1391–1398.

Laiho, R. and Prescott, C.E. (2004) Decay and nutrient dynamics of coarse woody debris in northern coniferous forests: a synthesis. Canadian Journal of Forest Research 34, 763–777.

Langor, D.W., Spence, J.R., Hammond, H.E.J., Jacobs, J., and Cobb, T.P. (2006) Maintaining saproxylic insects in Canada's extensively managed boreal forests: a review. In Grove, S. and Hanula, J. (eds) Insect Biodiversity and Dead Wood: Proceedings of a Symposium for the 22nd International Congress of Entomology. USDA Forest Service, Southern Research Station, Asheville, NC. General Technical Report GTR-SRS-93. pp. 83–97.

LeBarron, R.K. (1939) The role of forest fires in the reproduction of black spruce. Minnesota Academy of Science Proceedings 7, 11–14.

Lester, N.P., Dextrase, A.J., Kushneriuk, R.S., Rawson, M.R., and Ryan, P.A. (2004) Light and temperature: key factors affecting walleye abundance and production. Transactions of the American Fisheries Society 133, 588–605.

Lindgren, B.S. and Miller, D.R. (2002) Effect of verbenone on attraction of predatory and woodboring beetles (Coleoptera) to kairomones in lodgepole pine forests. Environmental Entomology 31, 766–773.

Lutz, H.J. (1956) Ecological effects of forest fires in the interior of Alaska. USDA Forest Service, Alaska Forest Research Center, Juneau, AK. Technical Bulletin 1133.

Manies, K.L., Harden, J.W., Bond-Lamberty, B.P., and O'Neill, K.P. (2005) Woody debris along an upland chronosequence in boreal Manitoba and its impact on long-term carbon storage. Canadian Journal of Forest Research 35, 472–482.

Maser, C., Trappe, J.M., and Nussbaum, R.A. (1978) Fungal-small mammal interrelationships with emphasis on Oregon coniferous forests. Ecology 59, 799–809.

Moore, R.D. and Richardson, J.S. (2012) Natural disturbance and forest management in riparian zones: comparison of effects at reach, catchment, and landscape scales. Freshwater Science 31, 239–247.

Morissette, J.L., Cobb, T.P., Brigham, R.M. and James, P.C. (2002) The response of boreal forest songbird communities to fire and post-fire harvesting. Canadian Journal of Forest Research 32, 2169–2183.

Morneault, C. and Payette, S. (1989) Postfire lichen–spruce woodland recovery at the limit of the boreal forest in northern Quebec. Canadian Journal of Botany 67, 2770–2782.

Muona, J. and Rutanen, I. (1994) The short-term impact of fire on the beetle fauna in boreal coniferous forest. Annales Zoologici Fennici 31, 109–121.

Murphy, E.C. and Lehnhausen, W.A. (1998) Density and foraging ecology of woodpeckers following a stand-replacement fire. Journal of Wildlife Management 62, 1359–1372.

Nappi, A. and Drapeau, P. (2009) Reproductive success of the black-backed woodpecker (*Picoides arcticus*) in burned boreal forests: are burns source habitats? Biological Conservation 142, 1381–1391.

Nappi, A. and Drapeau, P. (2011) Pre-fire forest conditions and fire severity as determinants of the quality of burned forests for deadwood-dependent species: the case of the black-backed woodpecker. Canadian Journal of Forest Research 41, 994–1003.

Nappi, A., Drapeau, P., Giroux, J.-F. *et al.* (2003) Snag use by foraging black-backed woodpeckers *(Picoides arcticus)* in a recently burned eastern boreal forest. Auk 120, 505–511.

Nappi, A., Drapeau, P., Saint-Germain, M., and Angers, V.A. (2010) Effect of fire severity on long-term occupancy of burned boreal conifer forests by saproxylic insects and wood-foraging birds. International Journal of Wildland Fire 19, 500–511.

Nappi, A., Drapeau, P., and Savard (2004) Salvage logging after wildfire in the boreal forest: Is it becoming a hot issue for wildfire? Forestry Chronicle 80, 67–74.

Nave, L.E., Vance, E.D., Swanston, C.W., and Curtis, P.S. (2011) Fire effects on temperate forest soil C and N storage. Ecological Applications 21, 1189–1201.

Nelson, J.L., Zavaleta, E.S., and Chapin, F.S. III, (2008) Boreal fire effects on subsistence resources in Alaska and adjacent Canada. Ecosystems 11, 156–171.

Niemi, G.J. (1978) Breeding birds of burned and unburned areas in northern Minnesota. The Loon 50, 73–84.

Noël, J. (2001) Régénération forestière après feu et coupe de recuperation dans le secteur de Val-Paradis, Abitibi. Thèse MSc, Université du Québec à Montréal, Université du Québec en Abitibi-Témiscamingue, Montréal, QC.

Ohmann, L.F. and Grigal, D.F. (1979) Early revegetation and nutrient dynamics following the 1971 Little Sioux forest fire in northeastern Minnesota. Forest Science 25(4 Suppl.).

Paine, T.D., Raffa, K.F., and Harrington, T.C. (1997) Interactions among scolytid bark beetles, their associated fungi, and live host conifers. Annual Review of Entomology 42, 179–206.

Pakkala, T., Hanski, I., and Tomppo, E. (2002) Spatial ecology of the three-toed woodpecker in managed forest landscapes. Silva Fennica 36, 279–288.

Pan, Y., Birdsey, R.A., Fang, J. *et al.* (2011) A large and persistent carbon sink in the world's forests. Science 333(6045), 988–993.

Paquin, P. (2008) Carabid beetle (Coleoptera: Carabidae) diversity in the black spruce succession of eastern Canada. Biological Conservation 141, 261–275.

Paragi, T.F., Johnson, W.N., Katnik, D.D. *et al.* (1996) Marten selection of postfire seres in the Alaskan taiga. Canadian Journal of Zoology 74, 2226–2237.

Perera, A.H., Dalziel, B.D., Buse, L.J., Routledge, R.G., and Brienesse, M. (2011) What happens to tree residuals in boreal forest fires and what causes the changes? Ontario Ministry of Natural Resources, Ontario Forest Research Institute, Sault Ste. Marie, ON. Forest Research Report No. 174.

Perera, A.H., Remmel, T.K., Buse, L.J., and Ouellette, M.R. (2009) An assessment of residual patches in boreal fire in relation to Ontario's policy directions for emulating natural forest disturbance. Ontario Ministry of Natural Resources, Ontario Forest Research Institute, Sault Ste. Marie, ON. Forest Research Information Paper No. 169.

Pettit, N.E. and Naiman, R.J. (2007) Fire in the riparian zone: Characteristics and ecological consequences. Ecosystems 10, 673–687.

Pierson, J.C. (2009) Genetic population structure and dispersal of two North American woodpeckers in ephemeral habitats. PhD Thesis, University of Montana, Missoula, MT. Available from http://avianscience.dbs.umt.edu/projects/documents/Pierson_BBWO_dissertation.pdf (accessed August 2012).

Pilliod, D.S., Bury, R.B., Hyde, E.J., Pearl, C.A., and Corn, P.S. (2003) Fire and amphibians in North America. Forest Ecology and Management 178, 163–181.

Pinel-Alloul, B., Prepas, E., Planas, D., Steedman, R., and Charette, T. (2002) Watershed impacts of logging and wildfire: Case studies in Canada. Lake and Reservoir Management 18, 307–318.

Planas, D., Desrosiers, M., Groulx, S.R., Paquin, S., and Carignan, R. (2000) Pelagic and benthic algal responses in eastern Canadian Boreal Shield lakes following harvesting and wildfires. Canadian Journal of Fisheries and Aquatic Sciences 57, 136–145.

Pohl, G., Langor, D., Klimaszewski, J., Work, T., and Paquin, P. (2008) Rove beetles (Coleoptera: Staphylinidae) in northern Nearctic forests. The Canadian Entomologist 140, 415–436.

Popescu, V.D. and Hunter, M.L. Jr, (2011) Clear-cutting affects habitat connectivity for a forest amphibian by decreasing permeability to juvenile movements. Ecological Applications 21, 1283–1295.

Rayner, A.D.M. and Boddy, L. (1988) Fungal Decomposition of Wood: Its Biology and Ecology. John Wiley & Sons, Ltd, New York, NY.

Richmond, H.A. and Lejeune, R.R. (1945) The deterioration of fire-killed white spruce by wood-boring insects in Northern Saskatchewan. Forestry Chronicle 21, 168–192.

Ripple, W.J. and Larsen, E.J. (2001) The role of postfire coarse woody debris in aspen regeneration. Western Journal of Applied Forestry 16, 61–64.

Rose, A.H. (1957) Some notes on the biology of Monochamus scutellatus (Say) (Coleoptera: Cerambycidae). The Canadian Entomologist 89, 547–553.

Ross, D.A. (1960) Damage by long-horned wood borers in fire-killed white spruce, central British Columbia. Forestry Chronicle 36, 355–361.

Rowe, J.S. (1983) Concepts of fire effects on plant individuals and species. In Wein, R.W. and MacLean, D.A. (eds) The Role of Fire in Northern Circumpolar Ecosystems. Scope 18. John Wiley & Sons, Ltd, New York, NY. pp. 135–154.

Rowe, J.S. and Scotter, G.W. (1973) Fire in the boreal forest. Quaternary Research 3, 444–464.

Ryoma, R. and Laaka-Lindberg, S. (2005) Bryophyte recolonization on burnt soil and logs. Scandinavian Journal of Forest Research 20(Suppl. 6), 5–16.

Saab, V.A., Dudley, J., and Thompson, W.L. (2004) Factors influencing occupancy of nest cavities in recently burned forests. The Condor 106, 20–36.

Saab, V.A., Powell, H.D. W., Kotliar, N.B. *et al.* (2005) Variation in fire regimes of the Rocky Mountains: implications for avian communities and fire management. Studies in Avian Biology 30, 76–96.

Saab, V.A., Russell, R.E., and Dudley, J.G. (2009) Nest-site selection by cavity-nesting birds in relation to postfire salvage logging. Forest Ecology and Management 257, 151–159.

Saint-Germain, M., Drapeau, P., and Buddle, C.M. (2007) Host-use patterns of saproxylic phloeophagous and xylophagous Coleoptera adults and larvae along the decay gradient in standing dead black spruce and aspen. Ecography 30, 737–748.

Saint-Germain, M., Drapeau, P., and Buddle, C.M. (2008) Persistence of pyrophilous insects in fire-driven boreal forests: population dynamics in burned and unburned habitats. Diversity and Distributions 14, 713–720.

Saint-Germain, M., Drapeau, P., and Hebert, C. (2004a) Comparison of Coleoptera assemblages from a recently burned and unburned black spruce forests of northeastern North America. Biological Conservation 118, 583–592.

Saint-Germain, M., Drapeau, P., and Hebert, C. (2004b) Landscape-scale habitat selection patterns of *Monochamus scutellatus* (Coleoptera: Cerambycidae) in a recently burned black spruce forest. Environmental Entomology 33, 1703–1710.

Saint-Germain, M., Drapeau, P., and Hebert, C. (2004c) Xylophagous insect species composition and patterns of substratum use on fire-killed black spruce in central Quebec. Canadian Journal of Forest Research 34, 677–685.

Saint-Germain, M. and Greene, D.F. (2009) Salvage logging in the boreal and cordilleran forests of Canada: Integrating industrial and ecological concerns in management plans. Forestry Chronicle 85, 120–132.

Saint-Germain, M., Larivee, M., Drapeau, P., Fahrig, L., and Buddle, C.M. (2005) Short-term response of ground beetles (Coleoptera: Carabidae) to fire and logging in a spruce-dominated boreal landscape. Forest Ecology and Management 212, 118–126.

Sauvard, D. (2004) General biology of bark beetles. In Lieutier, F.E.A. (ed.) Bark and Wood Boring Insects in Living Trees in Europe: A Synthesis. Kluwer Academic Publishers, London, UK. pp. 63–88.

Schieck, J. and Hobson, K.A. (2000) Bird communities associated with live residual tree patches within cut blocks and burned habitat in mixedwood boreal forests. Canadian Journal of Forest Research 30, 1281–1295.

Schindler, D.W., Newbury, R.W., Beaty, K.G., Prokopowich, J., Ruszcynski, T., and Dalton, J.A. (1980) Effects of a windstorm and forest fire on chemical losses from forested watersheds and on the quality of receiving streams. Canadian Journal of Fisheries and Aquatic Sciences 37, 328–334.

Schütz, S., Weissbecker, B., Hummel, H.E., Apel, K.H., and Schmitz, H. (1999) Insect antenna as a smoke detector. Nature 398(6725), 298–299.

Seip, D. and Parker, K. (1997) Use of wildlife tree patches by forest birds in the sub-boreal spruce (SBS) zone. British Columbia Ministry of Forests, Prince George, BC. Research Note #PG-08.

ECOLOGICAL ROLES OF WILDFIRE RESIDUALS

Simila, M., Kouki, J., Martikainen, P., and Uotila, A. (2002) Conservation of beetles in boreal pine forests: the effects of forest age and naturalness on species assemblages. Biological Conservation 106, 19–27.

Sirois, L. (1993) Impact of fire on *Picea mariana* and *Pinus banksiana* seedlings in subarctic lichen woodlands. Journal of Vegetation Science 4, 795–802.

Sirois, L. (1995) Initial phase of postfire forest regeneration in two lichen woodlands of northern Quebec. Ecoscience 2, 177–183.

Slaughter, K.W., Grigal, D.F., and Ohmann, L.F. (1998) Carbon storage in southern boreal forests following fire. Scandinavian Journal of Forest Research 13, 119–127.

Smith, T.M., Cramer, W.P., Dixon, R.K., Leemans, R., Neilson, R.P., and Solomon, A.M. (1993) The global terrestrial carbon-cycle. Water Air and Soil Pollution 70, 19–37.

St-Onge, I. and Magnan, P. (2000) Impact of logging and natural fires on fish communities of Laurentian Shield lakes. Canadian Journal of Fisheries and Aquatic Sciences 57(Suppl. 2), 165–174.

Stuart-Smith, K., Adams, I.T., and Larsen, K.W. (2002) Songbird communities in a pyrogenic habitat mosaic. International Journal of Wildland Fire 11, 75–84.

Taylor, D.L. and Barmore, W.J. (1980) Post-fire succession of avifauna in coniferous forests of Yellowstone and Grand Teton National Parks, Wyoming. In De Graf, R.M. and Tilghman, N.G. (eds) Management of western forests and grassland for non game birds. USDA Forest Service, Intermountain Research Station, Ogden, UT. General Technical Report GTR-INT-86. pp 130–140.

Thompson, I.D. and Anglestâm, P. (1999) Special species. In Hunter, M.L. Jr, (ed.) Maintaining Biodiversity in Managed Forests. Cambridge University Press, Cambridge, UK. pp. 434–459.

Toivanen, T. and Kotiaho, J.S. (2007) Mimicking natural disturbances of boreal forests: the effects of controlled burning and creating dead wood on beetle diversity. Biodiversity and Conservation 16, 3193–3211.

Tremblay, J.A., Ibarzabal, J., and Savard, J.L. (2009) Foraging ecology of black-backed woodpeckers *(Picoides arcticus)* in unburned eastern boreal forest stands. Canadian Journal of Forest Research 40, 991–999.

Treseder, K.K., Mack, M.C., and Cross, A. (2004) Relationships among fires, fungi, and soil dynamics in Alaskan Boreal Forests. Ecological Applications 14, 1826–1838.

van Balen, J.H. (1980) Population fluctuations in the great tit and feeding conditions in winter. Ardea 68, 143–164.

van Wilgenburg, S.L. and Hobson, K.A. (2008) Landscape-scale disturbance and boreal forest birds: Can large single-pass harvest approximate fires? Forest Ecology and Management 256, 136–146.

Viereck, L.A. (1983) The effects of fire in black spruce ecosystems of Alaska and northern Canada. In Wein, R.W. and MacLean, D.A. (eds) The Role of Fire in Northern Circumpolar Ecosystems. John Wiley & Sons, Ltd, New York, NY. pp. 201–220.

Viereck, L.A., Foote, J., Dyrness, C.T., van Cleve, K., Kane, D., and Siefert, R. (1979) Preliminary results of experimental fires in the black spruce type of interior Alaska. USDA Forest Service, Pacific Northwest Research Station, Portland, OR. Research Note PNW 332.

ECOLOGICAL ROLES OF WILDFIRE RESIDUALS

Villard, M. and Schieck, J. (1996) Immediate post-fire nesting by black-backed woodpeckers, *Picoides arcticus*, in northern Alberta. Canadian Field-Naturalist 111, 478–479.

Visser, S. (1995) Ectomycorrhizal fungal succession in jack pine stands following wildfire. New Phytologist 129, 389–401.

Visser, S. and Parkinson, D. (1999) Wildfire vs. clearcutting: impacts on ectomycorrhizal and decomposer fungi. In Meurisse, R.T., Ypsilantis, W.G., and Seybold, C. (eds) Proceedings of the Pacific Northwest Forest and Rangeland Soil Organism Symposium. USDA Forest Service, Pacific Northwest Research Station, Portland, OR. General Technical Report PNW-GTR-461. pp. 114–123.

Webb, E.T. (1998) Survival, persistence, and regeneration of the reindeer lichens, *Cladina stellaris*, *C. rangiferina*, and *C. mitis* following clearcut logging and forest fire in northwestern Ontario. Rangifer 10, 41–47.

Whelan, R.J. (1995) The Ecology of Fire. Cambridge Studies in Ecology, Cambridge University Press, Cambridge, UK.

Wikars, L.-O. (1997) Effects of forest fire and the ecology of fire-adapted insects. Acta Universitatis Upsaliensis, Uppsala, Sweden. Comprehensive Summaries of Uppsala Dissertations, Faculty of Science and Technology 272.

Wikars, L.-O. (2001) The wood-decaying fungus *Daldinia loculata* (Xylariaceae) as an indicator of fire-dependent insects. Ecological Bulletins 49, 263–268.

Wikars, L.-O. (2002) Dependence on fire in wood-living insects: An experiment with burned and unburned spruce and birch logs. Journal of Insect Conservation 6, 1–12.

Yunick, R.P. (1985) A review of recent irruptions of the black-backed woodpecker and 3-toed woodpecker in eastern North America. Journal of Field Ornithology 56, 138–152.

Zasada, J.C. and Gregory, R.A. (1969) Regeneration of white spruce with reference to interior Alaska: a literature review. USDA Forest Service, Pacific Northwest Forest and Range Experiment Station, Portland, OR. Research Paper PNW-79.

ECOLOGICAL ROLES OF WILDFIRE RESIDUALS

5 Role of wildfire residuals in forest management applications

A scrutiny of the recent scientific literature on boreal wildfire residuals will reveal that interest in this topic is not merely academic. Wildfire residuals have also become the focus of forest management applications in both Europe and North America where boreal forests occur, which is evident from the proliferation of research and the increase in published articles during the last two decades. Among these efforts, we recognize three different forest management applications that focus on boreal wildfire residuals.

The first, an attempt to restore wildfire residuals and downed wood by reintroducing fire to the landscape, is evolving in regions where wildfires have become infrequent as a result of land-use changes and aggressive fire suppression (Granström 2001, Wallenius 2011). Restoring wildfires is an emerging practice, particularly in Fennoscandia, where it is being discussed intensely among

Ecology of Wildfire Residuals in Boreal Forests, First Edition. Ajith H. Perera and Lisa J. Buse.
© 2014 Ajith H. Perera and Lisa J. Buse. Published 2014 by John Wiley & Sons, Ltd.
Companion Website: www.wiley.com/go/perera/wildfire

ecologists. The second involves the use of spatial and temporal patterns of residual patches and snag residuals as a template to design forest harvesting activities. This practice, termed "emulating natural forest disturbance", has gained popularity in boreal North America (Perera *et al.* 2004a). The third application is an approach to harvesting that aims to recover the standing timber from among the residual patches and snag residuals and that is becoming increasingly popular in North America, where boreal wildfires are common. However, this practice, termed "salvage logging", is controversial and has become the center of recent ecological debate (Lindenmayer *et al.* 2008).

Our goal in this chapter is to discuss these three applications by examining their origins, evolution, present state, and future prospects. In doing so, we draw strongly on the contents of the preceding chapter (the ecological roles of residuals), but do not repeat the specific details. Also, as we point out later, all of these applications have strong economic and social elements in addition to their ecological foundation, but a comprehensive discussion of these socio-economic dimensions is beyond the scope of this book. We also do not address or discuss several other forest management applications that involve, for example, curtailing the formation of residuals through fire suppression. In addition to reviewing and synthesizing the published information about the forest management applications, we provide examples of the practices, the development of related forest management policies and strategies, and the design and implementation of forest harvesting approaches, using examples from Fennoscandia and North America.

Restoring wildfire residuals

In contrast to the natural forest dynamics that continue to occur in much of the boreal forest in Alaska, Canada, and Russia, forest ecosystem processes in parts of Fennoscandia have been noticeably changed by human actions (Kuuluvainen *et al.* 2002). Specifically, wildfire occurrence in most of Fennoscandia has been almost completely eliminated, so that the annual burned area is reduced to less than 0.01% of the forested land (Granström 2001). This trend started at the beginning of the 20th century (Figure 5.1), and is generally attributed to changes in land use and societal values such as the adoption of modern agriculture and forestry (e.g., Parviainen 1996, Wallenius 2011) and the protection of people from natural hazards (Pyne 1996). The dramatic decrease in wildfire effects is specifically attributed to fire-suppression efforts (e.g., Granström 2001). It is widely asserted, however, that eliminating wildfires has resulted in a progressive reduction or loss of many fire-associated and fire-dependent species (e.g., Toivanen and Kotiaho 2007), causing concern among ecologists about the loss of biodiversity in Fennoscandian forest landscapes (e.g., Angelstäm 1998, Kuuluvainen 2002, Angelstäm *et al.* 2011).

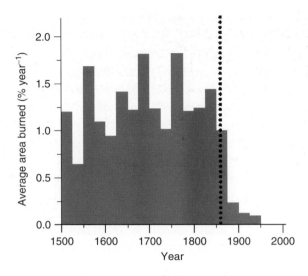

Figure 5.1   Wildfires decreased dramatically in central Fennoscandia following the implementation of fire-suppression policies. The graph shows the average annual area burned by wildfire during the past five centuries in central Fennoscandia; the onset of fire-suppression policies is indicated by the dotted vertical line. (Wallenius 2011. Reproduced with permission from T. Wallenius, University of Helsinki.)

As early as 1996, Pyne speculated that efforts to reintroduce fire to restore landscapes (i.e., to return these areas to more natural conditions so that their intrinsic ecological processes can resume and their biodiversity can recover) will likely increase in Fennoscandia. Since then many others also have recognized the need to restore these forest landscapes by reintroducing fires (Esseen *et al.* 1997, Angelstäm 1998, Granström 2001). In fact, a prominent scientific journal that focuses on forestry in Fennoscandia recently devoted a special issue to this topic [*Silva Fennica* 2011, 45(5)]. As we described in Chapter 2, given that boreal forests are fire-driven ecosystems, reintroducing fire in the form of controlled burning is viewed as one promising option to restore key ecological processes in Fennoscandian forests. For example, forest certification programs in Sweden include recommendations that forest managers actively use fire, with the goal of burning a given amount of the managed area every year (Ehnström 2001, Granström 2001). Despite the relatively small annual area burned in Finland's state-owned forests, fires have been applied as a restoration measure for nearly two decades.

Ecological expectations

In this context, *restoration* refers to measures taken in degraded forest ecosystems to recreate conditions that resemble the natural state (Similä and Junninen

2012), and is essentially a form of assisted ecosystem recovery. The goal is to create post-fire conditions that will lead to the restoration of the associated ecological processes. This includes diversifying the species composition by encouraging regeneration from residual vegetation after the fire; that is, sprouting from stems and roots of deciduous species and seeding-in of serotinous species from snag residuals (Similä and Junninen 2012). The increased species diversity is expected to recreate the early successional stages represented by the specific forest habitat types that result from wildfires. Also, restoring key stand structural attributes, such as snag residuals, mounds and pits caused by falling stems, a more open and thus warmer habitat, and burned substrates, will allow the species that depend on these features to recolonize (Kuuluvainen et al. 2002).

However, the primary intent of restoration efforts using fire is to create a reservoir of charred snags and downed wood; that is, the goal is to use fire to provide a continuous supply of decaying wood over time (Linder et al. 1998, Ehnström 2001). These snags and downed wood are expected to create habitat for fire-associated species, and particularly for saproxylic beetles (Ehnström 2001, Toivanen and Kotiaho 2007, 2010). As we discussed in Chapter 4, some of these species are considered either threatened or endangered. In some instances, creating dead wood by burning stands to provide habitat for pyrophilous beetles has been successful. For example, Toivanen and Kotiaho (2007, 2010) found that controlled burns in Norway spruce (Picea abies) forests increased the species diversity of pyrophilous beetles. They also found that burning facilitated the recovery of declining beetle species in the short term more successfully than simply creating dead wood in unburned areas. Similarly, Wikars (2002) determined that burned logs supported a higher diversity of beetle species than unburned logs, and that many pyrophilous species require large burns. In such cases, burning individual tree stems or placing burned logs in unburned areas is not a viable alternative to burning larger areas. The residuals created by fire evidently provide habitat features that are not available from dead wood created in other ways, evidence that underscores the importance of snag residuals and downed wood in these forests.

Considerations for application

Restoration template

Developing effective wildfire restoration strategies for boreal forests requires a consideration of what to restore. Since the target is a forest that resembles the "natural system", a clear understanding of that natural system is required from the outset. In Fennoscandia, as elsewhere, debate continues about the "natural" conditions that should be used as a template for the restoration efforts (Kuuluvainen

Figure 5.2 Ensuring that controlled burns create structural features and processes similar to those created by wildfire disturbances requires a thorough understanding of fire behavior in a range of forest types. Here, a restoration burn is underway in a spruce-dominated managed forest stand in Finland to create charred trees, snag residuals, and downed wood. (Reproduced with permission from Metla/Erkki Oksanen.)

2002). As Granström (2001) noted, controlled burning is typically implemented in previously thinned or partially harvested stands soon after a rain to make the fire easier to control, with the result that the fire has little effect on the soil; however, the fire has a high intensity due to the high fuel load, so tree mortality is high. This is the opposite of what would have happened during natural fires in boreal Europe and Fennoscandia, since natural fires typically let many trees survive, thereby providing long-term inputs of dead wood to the site, and since these fires also burned deep into the organic soils that are common in this region. Thus, the ecological effects of controlled burns will differ from those caused by a wildfire. Similä and Junninen (2012) suggested that, in controlled burns in Finland, some portion of the trees should be left alive after the fire, along with some unburned patches (e.g., in wet areas) to ensure that the effects more closely resemble those caused by a wildfire (Figure 5.2).

Spatiotemporal aspects

Given that the goal is to restore the effects of natural fire, restoration must be considered at many spatial and temporal scales. The strategy is to create a mosaic of stand structures and successional stages, with adequate representation and connectivity among stands at regional scales. At a broader scale, a connected network of burned forest areas can be developed that will facilitate the movement of some species. For example, Virkkala *et al.* (1993) proposed such a network of patches in Finland that would be large enough to provide adequate habitat for the endangered white-backed woodpecker (*Dendrocopos leucotos*). Originally a fire-dependent species, this woodpecker is thought to have survived in forests that originated from slash and burn cultivation, but to thrive it needs large (50–100 ha) areas of open deciduous forest that contain some dead wood. As a result of fire suppression, these areas have become fewer, and this woodpecker must now fly longer distances in search of suitable habitat.

In addition to considering the size and spatial distribution of burned areas, the recurrence of fire at a given location is also important. As Granström (2001) pointed out, mobile fire-dependent organisms (e.g., insects) can use the dispersed patches created by small wildfires in a landscape. Likewise, sessile organisms (e.g., fire-dependent plant species with seed banks) and soil processes that are improved by fire (e.g., the ameliorating effect of charcoal) benefit from shorter fire frequencies.

Kuuluvainen (2002) outlined a conceptual nested hierarchical approach that considers landscape, stand, and microsite scales and that was intended to create habitat characteristics similar to those that would result from the disturbance dynamics occurring in natural forests. Kuuluvainen *et al.* (2002) provided additional details of the specific characteristics to be considered at landscape and stand

Table 5.1 Processes and structural features characteristic of boreal forests that are the focus of restoration efforts involving fires in Fennoscandia (adapted from Kuuluvainen *et al.* 2002).

Scale	Processes to be restored	Structural features to be restored
Landscape	Fire regime Distribution and spatial pattern of successional stages Distribution and spatial pattern of fire-free areas (refugia) Dynamics and spatial distribution of dead wood	Size distribution of habitat patches Post-fire successional stages
Stand	Post-fire successional stages Multiple pathways of wood decay succession	Standing dead trees Fallen dead wood Fire-scarred trees, snags, and stumps Structurally and compositionally diverse understory canopies

ROLE OF WILDFIRE RESIDUALS IN FOREST MANAGEMENT APPLICATIONS

scales; Table 5.1 provides a summary of the details related to post-fire processes and residual structure.

Despite recognition of the importance of the landscape-level spatial and temporal considerations related to restoration efforts, much of the research and implementation remains focused at stand level, where restoration goals are often related to recreating one key aspect, such as the presence of large, old, dead trees in a small area. For example, controlled burning was used in an old-growth forest in Sweden to create large dead pines with open fire scars (Linder *et al.* 1998). In Finland, a more extensive restoration study included 24 stands, but the areas treated were each smaller than 3 ha (Vanha-Majamaa *et al.* 2007). Although such efforts may provide habitat for certain target species, they do not address regional-level concerns such as the distribution and connectivity of habitats (Niemi 1999, Angelstäm *et al.* 2011). However, some have argued that it is unrealistic to maintain whole landscapes with a natural fire mosaic in present-day Fennoscandia. They contend that a collection of small controlled burns should be sufficient to maintain most fire-dependent insect and avian species, since many have efficient long-distance dispersal mechanisms (e.g., Esseen *et al.* 1997).

Social context

Beyond a good understanding of the system to be restored, and knowledge of where, when, and how best to implement specific practices, the effective implementation of restoration efforts may require that existing forest policies to be reviewed and revised to be accepted by the public. For example, despite the emphasis on the need to increase the abundance of dead wood to provide habitat for saproxylic species, forest policy in Sweden has traditionally limited the amount of dead coniferous wood that can be left behind, even in burned areas, and leaving more requires special permission (Ehnström 2001). In addition to policy adjustments, other obstacles to the use of fire for restoration may include logistical factors, such as cost and feasibility, as well as social opposition to the use of fire.

Challenges and uncertainties

Sound knowledge of the target ecosystem state is a prerequisite for the effective design and implementation of restoration strategies. In an effort to restore wildfire residuals, it is the knowledge of boreal wildfires and their effect on the landscape that is required. In Fennoscandia, burning for conservation purposes has mostly been undertaken in previously managed stands. Some researchers have expressed a concern that, given the considerable variability in boreal fire-driven ecosystems, a much better understanding of disturbance and succession cycles and of the effects of fires on stand dynamics in natural forests at multiple scales is required before guidelines can be developed for using fire in managed systems (Linder *et al.* 1998, Kuuluvainen 2002, Kuuluvainen *et al.* 2002). Toivanen and Kotiaho (2007, 2010) contend that not only is the general effect of fire on species assemblages poorly understood but the understanding of the ecological effectiveness of restoration practices also remains limited. Such knowledge deficiencies exist with respect to the creation of fire residuals as well. Furthermore, restoration treatments are often focused on one aspect of a system in isolation, such as the creation of dead wood. Applying these treatments effectively at a landscape scale requires a more thorough understanding of the larger system, including interaction effects. For example, restoration actions may interfere with natural processes, causing unanticipated effects on future natural disturbances (Kuuluvainen 2002). The need to acquire more reliable knowledge of a highly variable and complex process such as wildfire, and then designing suitable restoration practices – how, when, and where to apply fire intentionally for maximum ecological effectiveness – poses a serious challenge.

Beyond the knowledge uncertainties and the practical difficulties of implementing wildfire-based restoration, concerns have been raised about its validity. Some researchers are concerned about the ecological effects of reintroducing fires on forest pests and diseases that proliferate under post-fire conditions (McCullough *et al.* 1998, Parker *et al.* 2006); others note the potential economic effects of damage to adjacent forests caused by increased populations of pyrophilic beetles that are attracted to snag residuals (Kuuluvainen *et al.* 2002). Public reluctance is another obstacle to restoring wildfires, and this has been seen in many parts of the world. For example, in the western United States, despite policies that encourage the use of fire for ecological restoration and the re-establishment of natural fire regimes (Omnibus Public Land Management Act 2009), the use of fire is restricted primarily to fuel reduction in an effort to avoid intense wildfires and is generally not applied to the re-creation of post-fire systems for ecological reasons. The reluctance to reinstate fire as one of the landscape's natural processes may be linked to the social perception of fire as a destructive force that devastates mature forests and wilderness and that is seen as a threat to life and property.

Emulating wildfire disturbance

Background

The second forest management practice involves applying knowledge of wildfire residuals to guide how boreal forests are harvested. This forest management paradigm – emulating natural disturbances – has emerged as a surrogate goal for

ROLE OF WILDFIRE RESIDUALS IN FOREST MANAGEMENT APPLICATIONS

forest sustainability in parallel with increasing recognition during the 1980s that periodic disturbance is an integral part of forest landscapes (Perera and Buse 2004). This approach was first proposed by Noss (1987) and Hunter (1990) as a practical means of applying the concept of a "coarse filter–fine filter" approach to the conservation of biodiversity. Emulating natural forest disturbance in the design of forest harvesting plans has garnered attention from ecologists worldwide, especially in the boreal regions of North America (e.g., Hunter 1993, Bergeron and Harvey 1997), Fennoscandia (e.g., Niemi 1999, Kuuluvainen 2002), and Asia (e.g., Liu *et al.* 2012). Readers will find many comprehensive publications that focus on the concepts that underlie this approach and their application. Examples include a special issue of a Finnish scientific journal [*Silva Fennica* 2002, 36(1)], a textbook (Perera *et al.* 2004a), and a literature review (Kuuluvainen and Grenfell 2012).

In essence, this approach attempts to mimic patterns – related to both structure and composition – that result from wildfires at a range of spatial scales during forest harvesting. However, many terms and definitions have been used in the literature to describe and define this practice, mostly related to specific goals and local circumstances. To alleviate the resulting confusion, we proposed the following global definition: "Emulating natural disturbance is an approach in which forest managers develop and apply specific management strategies and practices, at appropriate spatial and temporal scales, with the goal of producing forest ecosystems as structurally and functionally similar as possible to the ecosystems that would result from natural disturbances, and that incorporate the spatial, temporal, and random variability intrinsic to natural systems" (Perera and Buse 2004). Using this definition, the characteristics of wildfire disturbances can be emulated at appropriate scales during the design of forest harvesting plans (Perera and Cui 2010). At regional scales, wildfire disturbance can be emulated based on the characteristics of the regional fire regime. At sub-regional scales, emulation can be based on the characteristics of individual wildfires; that is, at the scale of harvest events on the geometry of wildfires and at the within-harvest-event scale on the compositional and spatial characteristics of wildfire residuals (Figure 5.3). For example, the forest harvest frequency, the extent of harvesting within a region, and the harvest block sizes can be based, respectively, on the regional fire-return interval, the total annual burned area, and the fire-size distribution. Similarly, the patterns of unburned forest patches and residual trees – both live and dead – within a wildfire footprint can be emulated during harvesting operations within a harvest footprint. It is the disturbance footprint that is the scale of our focus in the following section on the emulation of wildfire residuals during harvesting.

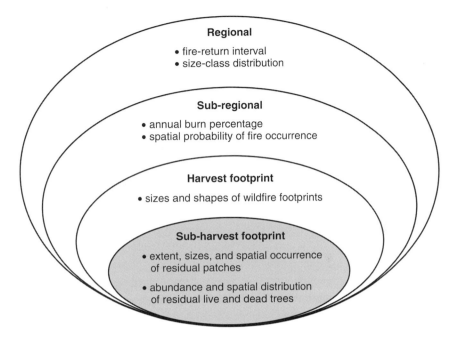

Figure 5.3   The characteristics of wildfires can be emulated by forest harvesting at several scales, nested within the regional and sub-regional scales. At the finest scale, patterns of wildfire residuals can be used to inform the design of timber harvest patterns within a harvest footprint.

Emulating wildfire residuals by forest harvest

The practice of emulating wildfire residual patterns has found its way into the policies and practices of several forest management agencies in boreal forest regions, and especially those in Canada. One example is the province of Ontario, where the need to emulate forest disturbances was embodied in legislation established during the early 1990s, and was incorporated into forest management policies and prescriptive guidelines for planning forest harvesting by 2000 (OMNR 2001). As a result of these changes, the emulation of natural disturbance has been implemented within forest management planning for more than a decade. Recently, the policies have been evaluated for their scientific validity, and were revised to form a second generation of directions (OMNR 2010). Several other provincial governments in Canada have also included the goal of emulating natural forest disturbance in their forest management strategies, such as British Columbia (BCMF 1995) and Alberta (ASRD 2008). The province of Québec is currently developing policy directions, which will soon be formally released. One reason for the increased popularity of this approach among public forest management agencies is the appeal of an approach to forest harvest that does not provide the aesthetically undesirable effects of clearcutting. In contrast to traditional clearcut harvests, which once dominated the managed boreal forest and left behind a landscape that appeared denuded and desolate, harvest footprints that emulate wildfires appear less barren and more structurally diverse (Figure 5.4).

Although the basic ecological premise of emulating natural disturbance is clear, and is based on the patterns of residuals within wildfire footprints, the exact prescriptions for characteristics that must be emulated by harvesting (i.e., the "emulation criteria") appear to vary regionally and among forest management agencies. Such prescriptive templates include emulation criteria for analogs of residual patches, live tree residuals, and snag residuals, all of which must be retained, not harvested, during logging activities within a harvest footprint (Table 5.2). The prescriptions mostly recommend single values for the minimum sizes of residual patches and the minimum abundance of trees and snags, instead of a range of values reflecting the range of abundances that would occur following fires (as discussed in Chapter 3). Some agencies (e.g., ASRD 2008) prescribe variable retention, in which retention levels are based on the area of the harvest footprint, to introduce some additional heterogeneity in the residual structure. The directions for spatial distribution of post-harvest residuals are expected to follow the kind of spatial biasing that is typically seen within wildfire footprints.

Obtaining a sufficiently detailed knowledge of the residual patterns from studies of boreal wildfires, preferably under the same forest conditions as those of the stands under management, is one major challenge in designing templates for emulating the natural disturbance pattern. As we discussed in Chapter 3, most studies that defined and quantified the wildfire residual structure did so through a specific lens, such as that of bird habitat, beetle dynamics, or nutrient cycling. As a result, comprehensive information on wildfire residuals *per se* is scarce. This lack of knowledge is further compounded by the broad range of variability that is inherent to boreal wildfires, which makes it difficult to describe modal conditions for the abundance and spatial distribution of residuals desired by forest managers, as well as the potential range of variability in these parameters. Another challenge, albeit a practical one associated with setting policy guidelines, is how to translate information about wildfire residuals into a set of emulation criteria that can be monitored and assessed for their policy compliance, ecological effects, and effectiveness in achieving the goals of the emulation.

Despite these challenges, the policies developed thus far have influenced timber harvesting in a vast area of Canada's boreal forest. For example, during the last decade in Ontario, nearly 1 000 000 ha of boreal forest have been harvested

ROLE OF WILDFIRE RESIDUALS IN FOREST MANAGEMENT APPLICATIONS

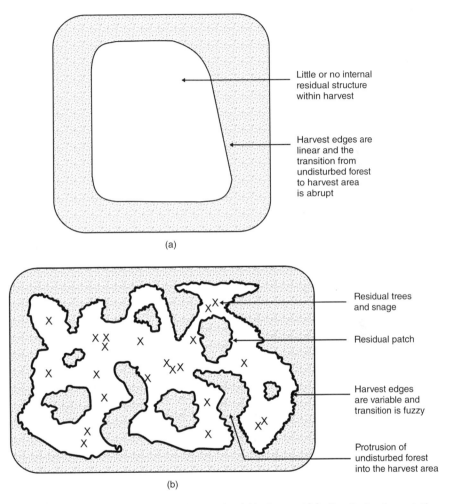

Little or no internal
residual structure
within harvest

Harvest edges are
linear and the
transition from
undisturbed forest
to harvest area
is abrupt

(a)

Residual trees
and snage

Residual patch

Harvest edges
are variable and
transition is fuzzy

Protrusion of
undisturbed forest
into the harvest area

(b)

Figure 5.4   Typical pattern of timber harvesting (a) before and (b) after the implementation of policies to emulate the pattern of wildfire residuals. Older harvests were linear, clearcut, and devoid of internal structure other than a few stems of no commercial value that were left behind. In contrast, timber harvests that emulate wildfire residuals are complex, with irregular boundaries, and they contain forest patches, trees, and snags that were retained to increase the structural diversity of the post-harvest landscape.

ROLE OF WILDFIRE RESIDUALS IN FOREST MANAGEMENT APPLICATIONS

Table 5.2 Prescriptions for emulating the characteristics of wildfire residuals in the guidelines established by some Canadian public forest management agencies. The emulation criteria specify the directions for creating analogues of wildfire residuals within a harvest footprint. Note: The emulation criteria presented here have been summarized for the sake of brevity. Please see the original documents for more detailed descriptions.

Post-harvest residual type	Emulation criteria (directions for retaining post-harvest residuals)	Location and source
Forest patches	2–8% of the harvested area; minimum size of 0.25 ha; patches representative of the original forest cover type; located within a harvest area similarly to how stands that escape a fire would be distributed	Ontario (OMNR 2001)
	In proportion to the harvest size: <20 ha: 4–40 small (<0.2 ha) patches; 20–60 ha: 20–60 small patches and up to 3 large (0.2 to 2 ha) patches; 60–100 ha: 60–100 small patches and 6–10 large patches; >100 ha: >100 small patches and 20 large patches Located near the edge of the harvested area to create a gradual transition in the forest cover and to minimize the windthrow potential; near ephemeral draws and intermittent streams	Alberta (ASRD 2008)
	Equivalent to the area harvested within a landscape unit: <40 ha for 10–20% of the landscape unit; 40–250 ha for 10–20% of the landscape unit; 250–1000 ha for 60–80% of the landscape unit	British Columbia (BCMF 1995)
Live trees and snags	Minimum of 25 live trees and snags per hectare; a minimum of 6 large-diameter live trees; well-spaced throughout the harvested area A range of tree species, diameters, and conditions (dead to healthy trees)	Ontario (OMNR 2001)
	Live trees at least 2 m tall located throughout the harvested area As many snags as feasible; snags must be >2 m tall	Alberta (ASRD 2008)
	Live and dead trees consistent with those that would escape fire	British Columbia (BCMF 1995)

based on the natural disturbance (wildfire) emulation policies. Furthermore, these guidelines have helped to create more diverse forest landscapes (Figure 5.5 and Figure 5.6) than were produced by the simple clearcuts of the past, where little if any residual structure was retained after the harvest.

Expectations and uncertainties

The ecological premise of emulating wildfire residual patterns (at local and broader scales) during timber harvesting in boreal regions is straightforward. The hypothesis is that boreal forest landscapes are naturally subject to frequent wildfires, and as a result, the forests are naturally resilient because their species assemblages have evolved to be adapted to the ecological processes created by this intrinsic disturbance. If the hypothesis is correct, then emulating wildfire disturbance patterns during timber harvesting will increase the likelihood for post-harvest recovery in the short term and sustainability of the ecosystems in the long term. In this reasoning, the evolution of ecosystem resilience under sustained natural disturbances is based on Holling (1973) and the variability of these disturbances is based on Landres *et al.* (1999). Since it is practically impossible to emulate all physical, chemical, and biological processes that occur during and after wildfires, managers use the post-fire residual structure as a surrogate for the consequences of these processes. As we discussed in previous chapters, the unburned areas and live tree and snag residuals can be emulated within a harvest footprint in the expectation that such patterns will lead to or support ecological processes, both terrestrial (e.g., Hunter 1993) and aquatic (e.g., Sibley *et al.* 2012), that are similar to those that occur naturally within a wildfire footprint. Based on their literature review, Kuuluvainen and Grenfell (2012) concluded that ecologists in general accept this argument.

Sustaining forest ecosystems must meet a certain holistic expectation; that is, the approach is not linked to enhancing only one or a few selected ecological values. Consequently, it is difficult to evaluate the success of the emulation. Most ecologists and forest managers have specific value-driven goals; these include improving the habitat for a selected species or species group, enhancing the regeneration of desired tree species, and decreasing sedimentation in bodies of water. All of these goals are relatively simple to monitor, evaluate, and assess. However, the general expectation is that creating analogs of post-fire residual structure will lead to ecological processes that will assist post-disturbance recovery and promote the likelihood of long-term sustainability. Realization of that result is not easily measurable.

Coincident with popularization of the natural disturbance emulation concept in North America, the provincial government of Ontario (Canada) enacted legislation in 1994 to emulate natural disturbances as a means to ensure the ecological sustainability of its forest landscapes. Its premise was the precautionary approach, following Aldo Leopold's conservation ethic, in which engineering of forest landscapes by timber harvesting is to be conducted only within the bounds of natural disturbances. Though an ecologically progressive and well-intended step, this act was subject to immediate and intense debate as a "leap of faith" and a "grand experiment" that critics believed lacked a solid scientific foundation.

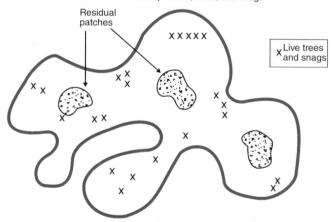

After lengthy scrutiny of the social, economic, and political feasibility of this approach, a forest policy document was released in 2001. This policy prescribed retaining forest patches, live trees, and snags during timber harvesting. Its specific directions for the extent and spatial distribution of residual patches and the abundance and distribution of live trees and snags were expected to emulate the residual structure produced by boreal wildfires. Over the last decade, nearly 1 million ha of timber harvests in boreal

(*continued*)
Ontario have been guided by this policy, creating a vast post-harvest landscape more diverse in structure than that produced by historical clearcuts.

As a part of Ontario's Environmental Assessment Act, the scientific foundation of the policy guide was recently examined through a series of studies. These revealed that the policy prescriptions were in general agreement with the characteristics of wildfire residuals. A new forest policy guide was released in 2010 that, in addition to the emulation wildfire residual patterns, placed more emphasis on retaining forest patches, live trees, and snags to provide specific habitat and environmental protection.

The immense variability that is inherent to wildfires and their post-disturbance patterns further complicates both the articulation of this goal and efforts to assess and evaluate its success. That variability and the gaps in scientific knowledge about how residuals, particularly unburned residual patches, are formed and spatially distributed within wildfire footprints, add to the difficulties in designing spatial and temporal emulation templates for forest harvest. Furthermore, the few scientific studies conducted on this practice have focused on individual species or species guilds (e.g., Hogberg *et al.* 2002, Harrison *et al.* 2005, Van Wilgenburg and Hobson 2008), as well as on selected ecological processes (e.g., Moore and Richardson 2012). As Kuuluvainen and Grenfell (2012) recently noted in their literature review, long-term comparisons of wildfires, clearcuts, and timber harvests that emulate natural disturbances are absent from the literature. This serious dearth of scientific evidence has caused some researchers to consider the concept of emulating natural disturbance to be an untestable hypothesis and in some instances even a leap of faith.

Expectations and uncertainties also extend to how practices designed to emulate the characteristics of wildfire residuals are implemented. Although the approach has been practiced for nearly two decades, its goals and expectations are not yet perceived clearly by the public and even by some forest managers. Despite cautions that this approach will not provide all ecosystem values and services in the short term, the emulation of disturbances has been both extolled as a panacea by its proponents and condemned as a futile exercise by its opponents (Perera *et al.* 2004b). Furthermore, only a subset of the potential spectrum of possible management options for wildfire emulation may be practiced due to social resistance based on aesthetics, conflicts with other values, or economic infeasibility (Sibley *et al.* 2012). Consequently, it is likely that use of the label "emulating wildfire residuals" may soon be discontinued even though this practice is gaining popularity worldwide because of its ability to diversify forest landscapes and provide a range of habitats (e.g., Gustafsson *et al.* 2012).

ROLE OF WILDFIRE RESIDUALS IN FOREST MANAGEMENT APPLICATIONS

(a)

(b)

Figure 5.5 (a, b) Examples of timber harvests (Ontario, Canada) designed to contain residual patches of forest that emulate the unburned patches left behind after boreal wildfires and create a more diverse landscape. This practice is becoming more common throughout boreal Canada, guided by specific forest policy directions implemented by public forest management agencies.

(a)

(b)

Figure 5.6 (a, b) Examples of timber harvests (Ontario, Canada) with live trees and snags remaining in the harvest block. The harvesting was intended to emulate the distribution and other characteristics of live tree and snag residuals after boreal wildfires. Retaining such post-harvest residuals is a standard practice throughout the boreal region, and it is expected to provide a diverse habitat that is more suitable for wildlife and other organisms.

ROLE OF WILDFIRE RESIDUALS IN FOREST MANAGEMENT APPLICATIONS

Salvage logging

The third forest management application related to wildfire residuals is contextually different from the two practices described above, which are motivated by the ecological value of the residuals in post-disturbance ecosystem processes. In contrast, the practice of harvesting wildfire residuals is motivated by economic reasons and directly involves manipulation of the aftermath of wildfires. In fact, the improved knowledge and wider recognition of the ecological values of the wildfire residuals may contribute to constraining the practice of salvage harvesting.

Background

Extraction of timber after wildfires has occurred in North America since the early 1900s (Peterson et al. 2009), but did not gain much popularity until recently, perhaps because access to wildfire locations was limited. However, this practice is now common worldwide, and also appears to occur frequently after other major forest disturbances such as insect infestations, wind storms, ice storms, and even volcanic eruptions (Lindenmayer et al. 2008). In boreal forest landscapes, harvesting timber in wildfire footprints has become a routine practice (Morissette et al. 2002, Schmiegelow et al. 2006, Saint-Germain and Greene 2009), and in some jurisdictions, where the occurrence of large wildfires is common, government agencies are developing policies to encourage this practice (Nappi et al. 2004, Schmiegelow et al. 2006). Concurrently, post-fire timber harvesting has become a recurrent topic in the scientific literature (e.g., McIver and Starr 2001, Morissette et al. 2002, Lindenmayer et al. 2004, Nappi et al. 2004, Dellasala et al. 2006, Donato et al. 2006, Lindenmayer and Noss 2006, Saint-Germain and Greene 2009, D'Amato et al. 2011). In addition, a special issue of a scientific journal [*Conservation Biology* 2006, 20(4)] and a textbook (Lindenmayer et al. 2008) have been focused on the ecological effects of harvesting wildfire residuals.

Readers who peruse the literature may note that many terms are used to describe the post-fire harvesting of timber. These include "clearcut logging after fire" (Purdon et al. 2004), "harvest of fire-killed trees" (McIver and Starr 2001), and "logging in burned forest" (Kotliar et al. 2002). However, the most commonly used term is "salvage logging". The exact nature of the harvesting may vary: the type of wildfire residuals harvested ranges from snags ("black trees") to unburned patches ("green trees"); the removal could include all or only a portion of the snag residuals; the areas harvested may be limited to a few selected parts of the wildfire or may include the whole footprint; and the harvesting period may range from the few weeks immediately after the fire to the first few years or even the first decade after the fire (Figure 5.7).

Figure 5.7   Typically, salvage logging involves extracting snag residuals from burned areas within a wildfire footprint. Harvesting commences almost immediately after the fire, as was the case in this example from a boreal wildfire in Ontario (Canada), where logging began within weeks after the fire was extinguished.

Regardless of the range of intensity and extent of the various harvesting activities, the appropriateness of removing wildfire residuals by salvage logging is widely questioned by ecologists. Many claim that after a severe natural disturbance, the forest is portrayed as being in a devastated condition, and that perception creates public indifference toward salvage logging (Lindenmayer and Noss 2006). In fact, salvage logging after fires and other natural disturbances is sometimes even publicly promoted as a remedial measure that will restore order to a devastated forest landscape and accelerate the recovery process (Figure 5.8). Some also contend that post-fire harvest was conducted to assist artificial regeneration or to reduce future fire hazards by reducing fuel availability. These are only ostensible reasons that justify the logging activities for economic goals (Lindenmayer and Noss 2006, D'Amato *et al.* 2011).

Whether or not the economic motivation is explicitly articulated, there is clearly some economic benefit from harvesting timber within wildfire footprints. In Canada, for example, many provincial government agencies provide economic incentives for timber salvage by levying the forestry company lower stumpage rates for the wildfire residuals (Schmiegelow *et al.* 2006). Partial or complete extraction of the wildfire residuals, both burned and unburned trees, is an

ROLE OF WILDFIRE RESIDUALS IN FOREST MANAGEMENT APPLICATIONS

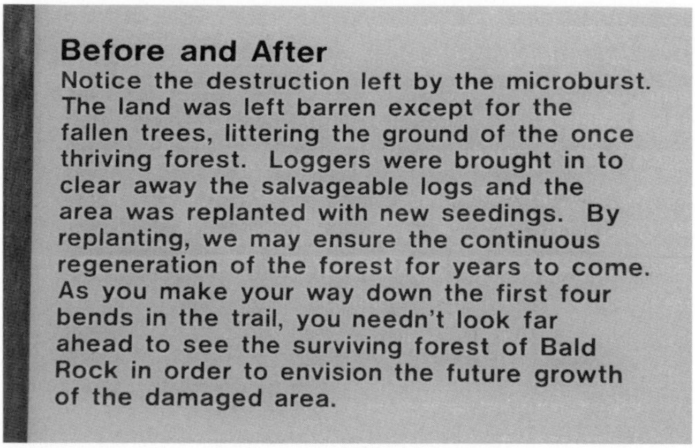

Before and After
Notice the destruction left by the microburst.
The land was left barren except for the
fallen trees, littering the ground of the once
thriving forest. Loggers were brought in to
clear away the salvageable logs and the
area was replanted with new seedings. By
replanting, we may ensure the continuous
regeneration of the forest for years to come.
As you make your way down the first four
bends in the trail, you needn't look far
ahead to see the surviving forest of Bald
Rock in order to envision the future growth
of the damaged area.

Figure 5.8   Natural forest disturbances have been portrayed as devastation, and their aftermath as aesthetically and ecologically undesirable. Salvage logging has been promoted as a means of restoring order in such cases. This photo shows a sign posted to justify salvage logging in a public forest after a natural disturbance (severe windthrow) that occurred in 2000 in Ontario (Canada).

opportunity for forestry companies to recover some of the economic losses caused by wildfires. For the forest industry, the motivation may be the recovery of lost revenue and investments such as in their road infrastructure; for government agencies, the motivation may be recovery of some of the lost tax revenues from timber harvest allocations. However, salvage logging is only feasible if there is convenient access to the burned area, and if the salvaged timber has a ready market. As a result, salvaging timber from wildfires commonly occurs near areas that were recently harvested or those with an established road network (Figure 5.9).

In the following sections, we examine the ecological considerations associated with salvage logging in detail and explore management practices that may mitigate the negative consequences. Although this discussion will cover some of the topics that we addressed in Chapter 4, we focus on how those ecological processes may be affected by salvage logging without repeating the details.

Ecological consequences

One general concern about salvage logging is the view that forest ecosystems are destroyed, damaged, or impaired following a major disturbance such as fire and that subsequent disturbances such as management interventions should, therefore, be avoided (Foster and Orwig 2006). Such secondary disturbance in an already

ROLE OF WILDFIRE RESIDUALS IN FOREST MANAGEMENT APPLICATIONS

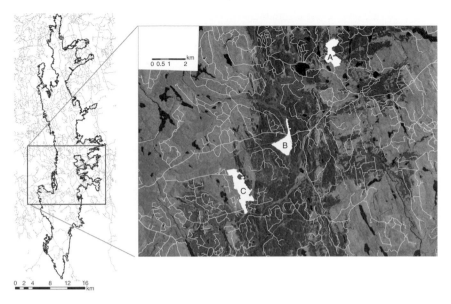

Figure 5.9   Extensive road networks and the proximity of markets for snag residuals make salvage logging feasible. For example, in this 40 000-ha wildfire in Ontario (Canada) (map), salvage logging of 300 ha (the areas marked A, B, and C in the inset photo, where the darker area represents the burned area) was expedited by the existing network of logging roads and a nearby market for salvaged logs.

disturbed ecosystem may indeed have cumulative deleterious effects (Lindenmayer *et al.* 2004, 2008). Many ecologists argue that salvage logging is detrimental to the recovering forest, the local environment, and even to regional biodiversity (Nappi *et al.* 2004, Dellasala *et al.* 2006, Donato *et al.* 2006, Lindenmayer and Noss 2006, Lindenmayer *et al.* 2008, D'Amato *et al.* 2011). These arguments address effects that extend beyond the usual concerns about the local ecological consequences of logging operations (McIver and Starr 2001, Karr *et al.* 2004).

The main factor contributing to negative ecological consequences appears to be the abrupt and extensive removal of snag residuals from the burned areas. A high abundance of snag residuals is the most prominent feature of a post-fire ecosystem. The snags constitute a very large pool of biomass, and one that will typically last for many decades, until the stem-exclusion stage of stand development (Figure 5.10). Sudden removal of the snag residuals and the mechanical operations associated with this removal are considered to be disruptive to ecosystem processes such as regeneration, species movement, hydrology, and nutrient cycling, and can also affect the recolonization dynamics (McIver and Starr 2001, Hebblewhite *et al.* 2009).

ROLE OF WILDFIRE RESIDUALS IN FOREST MANAGEMENT APPLICATIONS

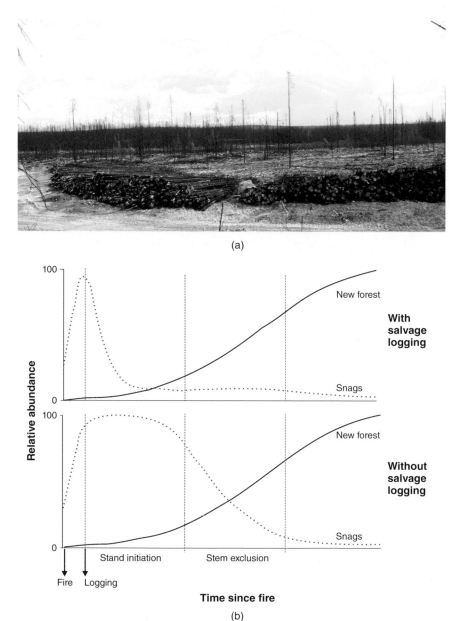

Figure 5.10   Salvage logging causes a sudden and early reduction in post-fire snag residuals. Harvested snag residuals may be removed from the site within a month of the wildfire (a; b, top). This contrasts with unharvested wildfire footprints, where snags may persist for decades (b, bottom). (Adapted from Peterson *et al.* 2009.)

Regeneration of vegetation

As we discussed in Chapter 4, the vast reservoir of snag residuals is the key driver of the resumption of ecological processes after the fire. The aerial seed bank that exists in the form of serotinous cones (e.g., jack pine, *Pinus banksiana*) and semi-serotinous cones (e.g., black spruce, *Picea mariana*) provides an on-site source of propagules that are essential for re-establishment of these species on burned sites (Figure 5.11). This is a particular concern for black spruce, which is slower to disperse its seed following fire than is jack pine (Greene *et al.* 2013). Immediate post-fire removal of snag residuals by salvage logging also removes the aerial seed source, leaving fewer seeds to fall on the site. Furthermore, the removal of snag residuals exposes the ground, changing the regeneration dynamics by favoring the suckering of broadleaved species such as aspen (*Populus* spp.), thereby altering the successional trajectory of the burned sites (Purdon *et al.* 2004, Greene *et al.* 2006, D'Amato *et al.* 2011). In addition to removal of the seed source, salvage logging removes the canopy cover, exposing the already scorched soil to sunlight, wind, and air movement, thereby changing the microclimate. For example, salvage logging could lead to drying of soil, resulting in species composition that reflects xeric conditions (Purdon *et al.* 2004). Poor regeneration of black spruce and dominance of aspen is one consequence of such drying conditions caused by salvage logging (Green *et al.* 2006). Also, composition of regenerating species in salvage-logged areas has been found to be more homogeneous than that in burned areas, possibly due to decreased microsite variability, increased soil temperatures, and the promotion of regeneration by species with reproductive tissues that are buried more deeply in the soil [e.g., aspen, alder (*Alnus* spp.)] and those with seed stored in the soil seed bank (Purdon *et al.* 2004, D'Amato *et al.* 2011).

Furthermore, salvage-logged areas appear to have a higher diversity of weedy species, such as graminoids and shrubs that prefer exposed sites (Kurulock and MacDonald 2007), but lower richness in typical understory species (Purdon *et al.* 2004). The increase in direct exposure, and subsequent drying of the soil, may be the cause of the decreased cover of mosses and other non-vascular species noted in salvage-logged areas (Kurulock and MacDonald 2007). In addition, salvage logging has been associated with decreased bryophyte abundance, especially for species that require burned and decaying wood as their substrate (Bradbury 2006). Delayed salvage logging, in which snag residuals are only removed after the regenerating seedlings and sprouts have become established, may impede ecosystem recovery by the damage caused to the regeneration by logging operations (Fraser *et al.* 2004, Greene *et al.* 2006). Beyond the removal of snag residuals, salvage logging sometimes involves the harvesting of residual patches within the burns, which may

ROLE OF WILDFIRE RESIDUALS IN FOREST MANAGEMENT APPLICATIONS

ROLE OF WILDFIRE RESIDUALS IN FOREST MANAGEMENT APPLICATIONS

(a)

(b)   (c)

Figure 5.11   (a) Jack pine cones (serotinous) and black spruce cones (semi-serotinous) on snag residuals provide a vast aerial seed bank to support natural regeneration after boreal wildfires, and these seeds can lead to rapid establishment of tree species: (b) jack pine seedlings 2 years after and (c) jack pine saplings 12 years after a fire.

decrease regeneration potential through the removal of this source of seed as well as through potential destruction of other propagules.

Beetle–woodpecker dynamics

Drawing on his research experience and an extensive review of the published literature, Hutto (2006) asserted that the effects of salvage logging differ from those that would occur following harvesting of an undisturbed forest. Salvage logging targets the same snags preferred by pyrophilous saproxylic beetles – large stems that are dying or dead, but only moderately charred – during the same period immediately after the fire. His review suggested that mass removal of snag residuals after the fire arrests the key ecological processes of beetle invasion followed by woodpecker invasion, both of which happen very rapidly after fires and occur at a vast scale, leading to a high abundance of beetles and woodpeckers (as we discussed in Chapter 4). Nappi and Drapeau (2009) are in agreement and suggested that a high density of newly formed snag residuals is crucial for the reproductive success of black-backed woodpeckers (*Picoides arcticus*). Hutto (2006) also posits that the effects of removing post-fire snag residuals differed from those of removing snags in an unburned forest due to differences in the avian species hosted by the burned stems. He further stated that in regions with a low fire frequency, the disruption of the beetle and woodpecker colonization processes would be a serious concern, and one that could also have ramifications for regional biodiversity.

   Many studies have been conducted to observe the avian species composition after salvage logging, sometimes in comparison with nearby sites that were not logged after a fire. The general consensus is that avian communities in salvage-logged areas differed from those found in burned areas that were left intact. The removal of perches, and of the food source represented by pyrophilous insects, led to decreases in numbers of timber-drilling and timber-gleaning species (Hutto and Gallo 2006). Indeed, Morissette *et al.* (2002) found that salvage logging reduced the number of resident bird species, canopy and cavity nesters, and insectivorous species. They argued that the resident avian species were displaced during logging operations, and that the others became less abundant because of a reduction in the number of stems suitable for nesting, perching, and feeding. In contrast, they also found that generalist bird species, omnivores, and ground- and shrub-nesting species were less affected by the logging. The bird species that were most sensitive to salvage logging appeared to be those associated with the early-successional post-fire forests, such as the three-toed (*Picoides tridactylus*), black-backed, and hairy (*Picoides villosus*) woodpeckers, the olive-sided flycatcher (*Contopus cooperi*), the house wren (*Troglodytes aedon*), the tree swallow (*Tachycineta bicolor*), and the northern flicker (*Colaptes auratus*) (Kotliar *et al.* 2002).

ROLE OF WILDFIRE RESIDUALS IN FOREST MANAGEMENT APPLICATIONS

The decline in bird species associated with partial or complete salvage logging was especially clear for post-burn specialist birds that use snag residuals immediately after a fire, such as the black-backed and three-toed woodpeckers (Hutto 2006). There is direct evidence that salvage logging decreased the total abundance of woodpeckers, and especially the black-backed woodpecker (Koivula and Schmiegelow 2007), and that these birds were most abundant in unsalvaged areas (Koivula and Schmiegelow 2007, Van Wilgenburg and Hobson 2008). In some instances, woodpeckers may be even completely absent in salvage-logged areas (Kotliar et al. 2002). This may be because larger stems are typically targeted by salvage logging, leaving mostly younger stands with smaller stems. As documented by Nappi and Drapeau (2009), such younger stands are unsuitable habitat for both saproxylic beetles and woodpeckers. Furthermore, they found black-backed woodpecker reproductive success to be lower in younger burned stands.

Other wildlife

The homogenizing effect on plant species composition caused by salvage logging (e.g., Purdon et al. 2004) has also been noted for ground beetles (Carabidae). Cobb et al. (2007) observed that the ground beetle assemblages became less diverse within salvage-logged sites where the pyrophilous ground beetle species associated with open habitats became more abundant during the first 2 years after the salvage logging (Cobb et al. 2007). Others have also found that exposure of a site by salvage logging favors ground beetles that prefer open habitats (e.g., Phillips et al. 2006). Conversely, the same open conditions created by salvage logging may adversely affect some species. For instance, these conditions decrease the availability of the high-moisture habitats and reproductive sites that are required by amphibians (Pilliod et al. 2003).

Exposure by salvage logging increases the sprouting of species such as aspen, and this can benefit wildlife that forage on this resource. However, the decreased snag cover renders foraging animals more vulnerable to predation, and the increased road network associated with salvage logging appears to provide increased access for predators. In a study conducted in boreal Alberta, Hebblewhite et al. (2009) noted that foragers such as elk (Cervus canadensis), deer (Odocoileus virginianus), and moose (Alces alces) avoided the use of salvage-logged stands, but continued to use unlogged burned areas. They estimated that the predation risk for such foragers was highest in salvage-logged areas.

Hydrology and nutrient cycling

Concerns about the ecological effects of logging operations in general (Figure 5.12), such as increases in soil erosion, excessive drying of soil, changes

(a)

(b)

Figure 5.12   (a) Skidding and (b) skid trails in a salvage logging operation after a boreal wildfire. Skid trails and the logging roads associated with salvage logging can increase soil erosion and sedimentation, and the removal of large snag residuals can decrease the continuous supply of inputs of coarse woody debris into streams.

in water flows, and increased sediment loads, also apply to salvage logging (McIver and Starr 2001, Karr *et al.* 2004). However, the negative consequences of salvage logging are considered to extend beyond those of standard logging. These consequences include channel alterations, sediment transport and deposition, and changes in water temperature (Minshall 2003, Reeves *et al.* 2006). Some researchers have observed increased sediment transport potential after salvage logging caused by timber-extraction roads and skid trails, and increased soil erosion potential caused by soil creep and the failure of stream banks (Silins *et al.* 2009). However, due to the high degree of variability associated with the degree of disturbance by wildfire, and variability in the intensity, extent, and timing of the salvage logging, it is difficult to generalize these effects. For example, post-fire salvage logging in moderately burned boreal forest areas decreased soil potassium, but in severely burned areas, soil magnesium and calcium also decreased (Brais *et al.* 2000).

Information about the effects of salvage logging in aquatic and riparian zones is lacking because no long-term monitoring programs exist, but researchers suspect that the removal of large snag residuals will decrease future inputs of coarse woody debris into stream channels, which in turn will negatively affect the fish habitat (Reeves *et al.* 2006). Similar concerns have also been voiced about the potential deleterious effects on benthic macroinvertebrates (Minshall 2003). According to Reeves *et al.* (2006), the removal of large snag residuals by salvage logging is also thought to increase surface soil erosion, because large wood typically traps fine sediments before they can be transported into streams. Furthermore, removing the snag residuals and surviving trees from riparian areas will reduce shading of the stream and could increase stream temperatures, especially if the harvesting operations affect the recovery of riparian vegetation that would shade the streams, thereby also affecting aquatic organisms.

Practical considerations

Given the economic realities that surround the practice of salvage logging, it may be unrealistic to expect that this practice will cease. In fact, with ever-expanding road networks that provide increasing access to forested areas, it is plausible that the incidence of salvage logging in wildfire footprints will increase. This is especially true if the occurrence of large wildfires increases in the boreal region, as some researchers predict will occur with changing climate, thereby diminishing the area available for conventional timber harvesting. Recognizing this, ecologists have offered many suggestions about how to modify salvage logging to minimize its negative ecological consequences. Some public forest management agencies have incorporated these suggestions into directions for salvage logging, as is the

Figure 5.13   Salvage logging often involves removing all the stems from a wildfire footprint, as in this example from Québec (Canada). (Reproduced with permission from J. Duval, Québec Ministry of Natural Resources and Wildlife.)

case for the province of Québec (Canada), where, in an extreme fire year, more than half of the wood volume harvested originated from salvage logging (Figure 5.13), an occurrence which led to the development of guidelines for ecosystem-based forest management (Nappi *et al.* 2011). Below we present a collection of recommendations that may moderate the ecological impacts of salvage logging.

Salvage only part of a wildfire footprint

Leaving a large portion of the wildfire footprint intact and unsalvaged is a common suggestion that is expected to allow natural ecological processes to progress unimpeded, particularly the invasion of pyrophilous beetles followed by woodpeckers (Hutto 1995, Kotliar *et al.* 2002). This suggestion is sensible, especially in areas where road construction and further disturbance would be required to extract salvageable timber. How much area should be left unharvested, and where these areas should be located within a wildfire footprint, may be guided by other factors such as the presence of ecologically sensitive areas and the requirement to supply critical habitat. For example, Nappi and Drapeau (2011) suggest that contiguous areas of burned mature and old forest larger than 20 ha be retained during post-fire harvesting to provide woodpecker habitat.

ROLE OF WILDFIRE RESIDUALS IN FOREST MANAGEMENT APPLICATIONS

Québec's Sustainable Forest Development Act allows salvage harvesting within burned areas under special exemptions from regular forestry practices in the province. The act minimizes economic losses by ensuring that burned timber is harvested swiftly, sometimes even without public consultation. Associated directives are intended to increase the rate of salvage logging following wildfires in Québec. For example, more than half of the wood volume harvested in one region in an extreme fire year came from salvage logging, and amounted to 20% of the total wood volume harvested in the province.

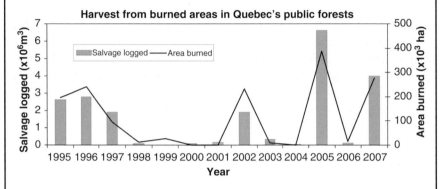

Recognizing the ecological importance of wildfire residuals, the Québec Ministère des Ressources Naturelles et de la Faune has adopted guidelines that will direct salvage logging practices to conserve biodiversity while extracting timber from wildfires. Consequently, an extensive literature review was conducted by an expert panel and a comprehensive set of management recommendations were made.

These recommendations embed salvage logging practices in the broader context of an ecosystem-based approach to managing Québec's forests, and emphasize proactive development of harvesting strategies rather than merely reacting to wildfire events. They recognize two scales to developing these strategies: the landscape scale (for many large fire events) and the fire scale (for a single fire event). At both scales, they specify the minimum extents of wildfires that must be left unharvested. At the scale of a single fire event, they recommend retaining a diverse collection of residual stands that are representative of the stand's pre-burn structure, composition, and distribution; prescribe the minimum sizes for residual patches; and recommend leaving riparian buffers and protecting other environmentally sensitive areas. It is particularly noteworthy that they recommend preserving all

(*continued*)
unburned forest within the wildfire footprint during salvage logging. Furthermore, they consider the option of delaying salvage logging whenever feasible.

At present, these recommendations are among the first formal efforts by a public forest management agency to merge the ecological values of wildfire residuals into salvage-logging practices.

Koivula and Schmiegelow (2007) also recommend leaving large clusters of snag residuals with the intent of countering the contraction in patch size that results from windthrow of such snags, which is more frequent along the edges. They also suggest that contiguous areas should be large enough to include the home ranges of birds that depend on dead wood, and should capture a range of species compositions, including both pure stands and mixedwoods with various levels of fire severity. Nappi *et al.* (2011) recommend leaving a minimum of 15% of each wildfire footprint unsalvaged, with the total unsalvaged area within a forest management unit in Québec (Canada) representing a minimum of 30% of the cumulative burned area over a 5-year period. Furthermore, they suggest not salvaging riparian buffers, unburned areas, and localities susceptible to soil erosion, as well as harvesting in strips and blocks to leave parts of the burned area intact. Saab *et al.* (2009, 2011) and Nappi *et al.* (2011) recommend maintaining areas of large snag residuals in wildfire footprints to maintain breeding habitats for cavity-nesting birds. Leaving such unlogged reserves in the center of wildfire footprints will promote the survival of some cavity-nesting bird species by providing habitat that is farther from predators that reside in the unburned forest (Saab *et al.* 2011). Another consideration is to ensure that the areas left unsalvaged reflect the pre-burn forest composition and structure and thus provide a range of habitats. This would prevent salvage harvesting of only mature forest with larger stems of preferred species, and leaving the younger forest and other areas that are deemed not commercially valuable (Nappi *et al.* 2011).

Salvage snag residuals selectively

One general recommendation is not to harvest the largest snag residuals (Kotliar *et al.* 2002, Nappi and Drapeau 2011). Woodpeckers prefer large snags for nesting (>20 cm diameter) and foraging (>15 cm diameter), especially when these are only moderately burned and occur at high densities (Nappi and Drapeau 2011). Another recommendation is to retain a variety of tree sizes and species. For example, to provide suitable woodpecker habitat, harvesting should leave broadleaved snag residuals for nesting, coniferous snags for foraging, a range of sizes to account for different preferences for foraging and nesting, and a range of stages of decay, since strong excavators prefer intact snags and other species prefer more decayed snags (Kotliar *et al.* 2002). Other researchers suggest retaining snag residuals partially to ensure the re-establishment of plants and colonization by other species. For example, Greene *et al.* (2006) recommend leaving a minimum of 10% of the large coniferous snag residuals as seed sources within the wildfire footprint, and Bradbury (2006) advises retaining snag residuals in a range of decay classes, size classes, and species groups to provide the range of woody substrates required to sustain a healthy bryophyte community. Nappi *et al.* (2011) recommend selectively leaving snag residuals unsalvaged, in proportion to the pre-burn compositional and structural diversity within a wildfire footprint.

Delay salvage logging

In addition to suggestions for modifying the extent and intensity of logging after wildfires, some researchers propose delaying the salvaging operations for several years (e.g., Murphy and Lehnhausen 1998). The premise of this proposal lies in the critical nature of the first few years after a fire for many ecological processes, including regeneration, beetle and woodpecker invasions, and nutrient cycling

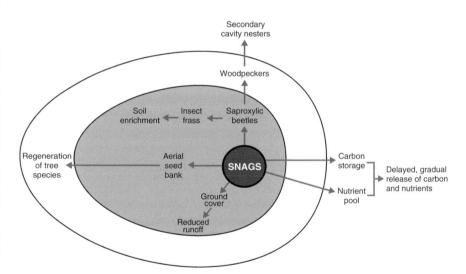

Figure 5.14   Delaying salvage logging by a few years can allow many ecological processes to continue unimpeded among the wildfire residuals. When snags are removed immediately after a fire all ecological processes outside the darkest shaded area are arrested. If salvage is delayed by as little as 2 years, as represented by the lighter shaded area, many ecological processes can proceed. A further delay in salvage logging, indicated by the outermost boundary, affects only long term-processes, thereby allowing the post-fire system to recover more rapidly.

(Figure 5.14). Delaying salvage logging could ensure that some of these processes will continue undiminished for a given period. However, in most cases the same time period is critical for the profitability of the harvest, and thus its economic feasibility. As Saint-Germain and Greene (2009) noted in their excellent review of the economic and ecological concerns about salvage logging, any delay in extracting snag residuals after a fire will result in significant economic losses. The degradation of wood due to boring by beetles, staining and decay by fungi, and checking (cracking of wood due to drying) sets in almost immediately after the fire, resulting in a rapid loss of timber quality and merchantable volume. This confluence of the critical time period for two competing concerns – ecological and economic – represents a serious practical obstacle to delaying salvage logging in wildfire footprints. However, it may be possible to alleviate this problem by exploring non-traditional avenues to utilize insect-damaged wood.

Plan at multiple scales

Although ecologists are concerned about the effects of salvage logging on regional biodiversity (e.g., Nappi *et al.* 2004, Hutto 2006, Lindenmayer and Noss 2006), most remedial solutions offered do not address the issue of scale. We suggest that both forest policymakers who develop salvage logging strategies and forest managers who plan the harvest operations that are designed under those strategies adopt a top-down approach (Figure 5.15). They must first consider the broader spatial contexts, such as the spatiotemporal frequency of wildfires (Figure 5.16), because wildfire footprints are part of a regional ecological process for which individual fires represent a single point in time. Next, they must consider the unburned matrix surrounding the wildfire footprint, to account for ecological processes that are initiated within the footprint but the effects of which extend beyond it (and vice versa). Finally, they must proceed to finer local scales to account for ecological processes that are restricted to the wildfire footprint and that begin and end within the fire boundary. In addition, by embracing an adaptive management approach to salvage logging at all scales, forest managers can progressively minimize the negative consequences of this practice (McIver and Starr 2001, Hutto 2006, Schmiegelow *et al.* 2006).

Uncertainties

Many uncertainties surround the appropriateness of salvage logging wildfire residuals from an ecological point of view. Perhaps as a result of these uncertainties, the existing policies that guide this practice are ambiguous and ambivalent (Schmiegelow *et al.* 2006, Saint-Germain and Greene 2009). Hutto

ROLE OF WILDFIRE RESIDUALS IN FOREST MANAGEMENT APPLICATIONS

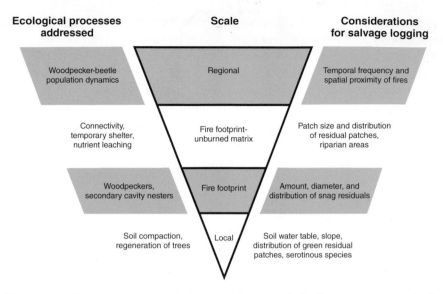

Ecological processes addressed	Scale	Considerations for salvage logging
Woodpecker-beetle population dynamics	Regional	Temporal frequency and spatial proximity of fires
Connectivity, temporary shelter, nutrient leaching	Fire footprint-unburned matrix	Patch size and distribution of residual patches, riparian areas
Woodpeckers, secondary cavity nesters	Fire footprint	Amount, diameter, and distribution of snag residuals
Soil compaction, regeneration of trees	Local	Soil water table, slope, distribution of green residual patches, serotinous species

Figure 5.15   A top-down approach that considers ecological effects at various scales is needed when developing policies and operational practices related to salvage logging. By adopting a hierarchical multiple-scale approach, coupled with adaptive management strategies, forest managers and policymakers may be able to reduce the negative consequences of salvage logging.

(2006) stated that the policies are grossly inadequate, and that many forest management agencies in North America actually have policies and taxing systems that encourage and enable salvage logging, while also expediting harvesting by providing minimal directions and guidelines. This is a concern that has been repeated by other researchers (Nappi *et al.* 2004, Lindenmayer and Noss 2006, Schmiegelow *et al.* 2006). Moreover, some ecologists (e.g., Lindenmayer and Noss 2006, Schmiegelow *et al.* 2006) have argued that public apathy exists towards salvage logging because burned forests are deemed to be already disturbed and unaesthetic, and therefore less valuable. This sentiment leads to a general acceptance of salvage logging, without any doubts about its ecological validity, and leads to a sense that little modification of natural resource management policies is required. However, the general consensus among ecologists (e.g., McIver and Starr 2001, Lindenmayer *et al.* 2004, Nappi *et al.* 2004, Hutto 2006, Lindenmayer and Noss 2006, Schmiegelow *et al.* 2006) appears to be that a set of ecologically appropriate policies is required to guide salvage logging, along with built-in monitoring programs that are implemented within an adaptive management framework (so that the results of monitoring can lead to modifications of the overall approach). Such a refinement of policies must be based on sound scientific knowledge, and must include a balanced and comprehensive appraisal of the existing ecological knowledge.

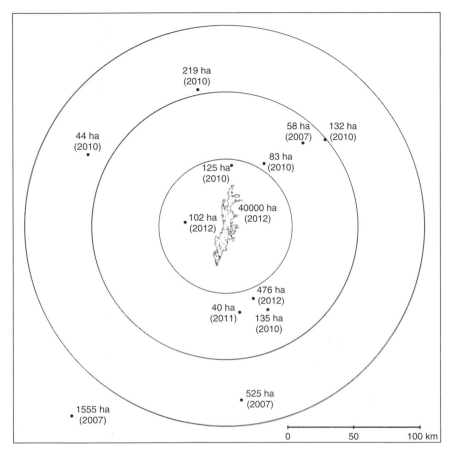

Figure 5.16 When considering the implementation of salvage logging, the regional disturbance context is important. As an example, locations of fires (dots; with extent and year of burn) that burned within a 150 km radius (concentric circles are at 50 km intervals) in the 5-year period before a 40 000-ha fire (center) indicate the distribution/availability of burned habitat in the region.

Beyond the uncertainties in the ecological knowledge about wildfire residuals that we outlined in Chapter 4, very little is known about the direct effects of a secondary disturbance such as salvage logging within burned areas or about the degree to which the ensuing ecological processes are affected (Saint-Germain and Greene 2009). These uncertainties are greater with respect to broad-scale ecological effects, for both landscape- and regional-level ecological processes, such as the preservation of regional continuity in the chain of pyrophilous beetles, woodpeckers, and secondary cavity nesters. These uncertainties are also high for the cumulative effects at regional scales, both spatially and temporally, such as the downstream effects of harvesting burned or unburned riparian residuals, removal of temporary refugia, and a failure to maintain ecological flows between burned and unburned areas. Although there is generally a better understanding of the effects of site-level factors such as soil compaction and the removal of the aerial seed bank on the regeneration of tree species, it is unclear how these factors affect processes at broader scales.

Summary

For close to two decades, wildfire residuals have been the focus of several management applications in boreal forests. This focus reflects two divergent perspectives: one is based on the recognition of the ecological importance of wildfire residuals and the other is based on the economic value of these trees. Their ecological value is the premise for creating wildfire residuals by reintroducing fire to restore natural ecological processes in landscapes where fire has effectively been eliminated, and to emulate the effects of natural disturbance by using the characteristics of wildfire residuals as a template for planning boreal timber harvests. Swift extraction of the burned and unburned tree residuals from wildfire footprints, referred to as "salvage logging", is primarily motivated by the commercial value of the timber.

In Fennoscandia, where changes in land-use patterns and aggressive fire suppression have all but eliminated wildfire from the landscape, the creation of wildfire residuals, and specifically charred dead wood, through restoring fire is seen as a way to provide critical habitat for declining fire-dependent species, thereby maintaining biodiversity. Although some success has been achieved for specific target species or species groups, many knowledge uncertainties remain; furthermore, several challenges, including social opposition to the reintroduction of wildfire, currently limit the widespread implementation of restoration efforts using fire.

Using wildfire residual patterns to design prescriptions for boreal timber harvesting has become a replacement for the traditional practice of clearcutting. This approach was conceived as a means to apply a coarse-filter approach to conserving biodiversity during large-scale commercial timber harvesting in boreal regions. This practice has rapidly gained popularity in boreal North America over the last two decades, and particularly in Canada, and has shaped many timber harvests to produce footprints that are more structurally diverse than those produced by clearcutting, affecting a combined area that may exceed millions of hectares. Proponents of emulating natural disturbances contend that, even though this is not a panacea, it is a precautionary approach in the absence of perfect and complete knowledge of the ecological impacts of large and continuous timber harvesting, especially in regions that are frequently disturbed by wildfires.

Their hypothesis is that when timber harvest patterns, including the characteristics of the post-harvest residuals, are designed to mimic the disturbance produced by wildfires, the resulting forest landscape will be more resilient and will conserve biodiversity more effectively. Unfortunately, these hypotheses have not been tested adequately and no consistent effort has been made to monitor the long-term ecological effects and effectiveness of this approach. Scientific and government enthusiasm for this practice may be waning because of an absence of sufficient scientific knowledge, which is exacerbated by economic arguments from opponents and by debates over the practical feasibility of implementing emulation templates – and perhaps also by inadequate communication between policymakers and the public. Such a shift may lead to an unfortunate abandonment of the safety provided by a coarse-filter approach in favor of retaining pre-selected focal values during boreal timber harvesting, which continues to occur at vast scales, with no assurance of overall forest sustainability.

Conversely, salvage logging meets an economic need, and is prompted by the desire to minimize economic losses by the commercial forest industry after a fire. This arises from the need to recover investments made in roads and other infrastructure in an area allocated for harvest under forest management plans, to sustain the supply of wood to processing centers, and to mitigate the loss of a forest company's annual harvest allocation. Salvage logging is therefore becoming increasingly common, and is likely to gain further momentum in the boreal forest as access to burned sites increases and as the available wood supply declines. The merits of salvage logging are presently heavily debated by ecologists. As we noted in this chapter, there are both local and regional ecological impacts that must be considered, and many government agencies lack comprehensive policies and directives to guide salvage logging. Many spatial (where to conduct salvage logging and how much to remove) and temporal (how long after fire) questions must be resolved. Reaching a resolution will be difficult, because many uncertainties exist about the post-fire ecological processes and the impacts of logging on these processes; moreover, the variability of post-fire residual structures and processes is high, and salvage logging itself is highly variable in its characteristics. Much of the specific information from studies in different geographic areas may not be of direct use in boreal forests because of differences in terrain, species assemblages, and fire regimes. Nonetheless, this information will be useful for developing research and management hypotheses. Much research will be needed to resolve the uncertainties, and proper policies and guidelines, together with post-harvest monitoring programs whose results provide feedback to management plans, will need to be instituted if this harvesting is to continue, given the associated ecological concerns.

ROLE OF WILDFIRE RESIDUALS IN FOREST MANAGEMENT APPLICATIONS

References

Angelstâm, P.K. (1998) Maintaining and restoring biodiversity in European boreal forests by developing natural disturbance regimes. Journal of Vegetation Science 9, 593–602.

Angelstâm, P., Andersson, K., Axelsson, R., Elbakidze, M., Jonsson, B.G., and Roberge, J.M. (2011) Protecting forest areas for biodiversity in Sweden 1991–2010: the policy implementation process and outcomes on the ground. Silva Fennica 45, 1111–1133.

ASRD (2008) Alberta sustainable timber harvest planning and operational ground rules framework for renewal. Alberta Sustainable Resource Development, Public Lands and Forests Division, Forest Management Branch, Edmonton, AB.

BCMF (1995) Biodiversity guidebook. British Columbia Ministry of Forests, Vancouver, BC.

Bergeron, Y. and Harvey, B. (1997) Basing silviculture on natural ecosystem dynamics: an approach applied to the southern boreal mixedwood forest of Quebec. Forest Ecology and Management 92, 235–242.

Bradbury, S.M. (2006) Response of the post-fire bryophyte community to salvage logging in boreal mixedwood forests of northeastern Alberta, Canada. Forest Ecology and Management 234, 313–322.

Brais, S., David, P., and Ouimet, R. (2000) Impacts of wild fire severity and salvage harvesting on the nutrient balance of jack pine and black spruce boreal stands. Forest Ecology and Management 137, 231–243.

Cobb, T.P., Langor, D.W., and Spence, J.R. (2007). Biodiversity and multiple disturbances: boreal forest ground beetle (Coleoptera: Carabidae) responses to wildfire, harvesting, and herbicide. Canadian Journal of Forest Research 37, 1310–1323.

D'Amato, A.W., Fraver, S., Palik, B.J., Bradford, J.B., and Patty, L. (2011) Singular and interactive effects of blowdown, salvage logging, and wildfire in sub-boreal pine systems. Forest Ecology and Management 262, 2070–2078.

Dellasala, D.A., Karr, J.R., Schoennagel, T. et al. (2006) Post-fire logging debate ignores many issues. Science 314: 51.

Donato, D.C., Fontaine, J.B., Campbell, J.L., Robinson, W.D., Kauffman, J.B., and Law, B.E. (2006) Post-wildfire logging hinders regeneration and increases fire risk. Science 311, 352.

Ehnström, B. (2001) leaving dead wood for insects in boreal forests – suggestions for the future. Scandinavian Journal of Forest Research 16(Suppl. 3), 91–98.

Esseen, P.-A., Ehnström, B., Ericson, L., and Sjöberg, K. (1997) Boreal forests. Ecological Bulletins 46, 16–47.

Foster, D.R. and Orwig, D.A. (2006) Preemptive and salvage harvesting of New England forests: when doing nothing is a viable alternative. Conservation Biology 20, 959–970.

Fraser, E., Landhausser, S., and Lieffers, V. 2004. The effect of fire severity and salvage logging traffic on regeneration and early growth of aspen suckers in north-central Alberta. Forestry Chronicle 80, 251–256.

Granström, A. (2001) Fire management for biodiversity in the European boreal forest. Scandinavian Journal of Forest Research 16(Suppl. 3), 62–69.

Greene, D.F., Gauthier, S., Noel, J., Rousseau, M., and Bergeron, Y. (2006) A field experiment to determine the effect of post-fire salvage on seedbeds and tree regeneration. Frontiers in Ecology and the Environment 4, 69–74.

Greene, D.F., Splawinski, T.B., Gauthier, S., and Bergeron, Y. (2013) Seed abscission schedules and the timing of post-fire salvage logging. Forest Ecology and Management 303, 20–24.

Gustafsson, L., Baker, S.C., and Bauhus, J. (2012) Retention forestry to maintain multifunctional forests: A world perspective. BioScience 62, 633–645.

Harrison, R.B., Schmiegelow, F., and Naidoo, R. (2005) Stand-level response of breeding forest songbirds to multiple levels of partial-cut harvest in four boreal forest types. Canadian Journal of Forest Research 35, 1553–1567.

Hebblewhite, M., Munro, R.H., and Merrill, E.H. (2009) Trophic consequences of postfire logging in a wolf–ungulate system. Forest Ecology and Management 257, 1053–1062.

Hogberg, L.K., Patriquin, K.J., and Barclay, R.M.R. (2002) Use by bats of patches of residual trees in logged areas of the boreal forest. American Midland Naturalist 148(2), 282–288.

Holling, C.S. (1973) Resilience and stability of ecological systems. Annual Review of Ecology and Systematics 4, 1–23.

Hunter, M.L. (1990) Wildlife, Forests, and Forestry: Principles of Managing Forests for Biodiversity. Prentice-Hall, Englewood Cliffs, NJ.

Hunter, M.L. (1993) Natural fire regimes as spatial models for managing boreal forests. Biological Conservation 65, 115–120.

Hutto, R.L. (1995) Composition of bird communities following stand-replacement fires in Northern Rocky Mountain (U.S.A.) conifer forests. Conservation Biology 9, 1041–1058.

Hutto, R.L. (2006) Toward meaningful snag-management guidelines for postfire salvage logging in North American conifer forests. Conservation Biology 20, 984–993.

Hutto, R.L. and Gallo, S.M. (2006) The effects of postfire salvage logging on cavity-nesting birds. The Condor 108, 817–831.

Karr, J.R., Rhodes, J.J., Minshall, G.W. et al. (2004) The effects of postfire salvage logging on aquatic ecosystems in the American West. BioScience 54, 1029–1033.

Koivula, M.J. and Schmiegelow, F.K.A. (2007) Boreal woodpecker assemblages in recently burned forested landscapes in Alberta, Canada: effects of post-fire harvesting and burn severity. Forest Ecology and Management 242, 606–618.

Kotliar, N.B., Hejl, S.J., Hutto, R.L., Saab, V.A., Melcher, C.P., and McFadzen, M.E. (2002) Effects of fire and post-fire salvage logging on avian communities in conifer-dominated forests of the western United States. Studies in Avian Biology 25, 49–64.

Kurulok, S.E. and MacDonald, S.E. (2007). Impacts of postfire salvage logging on understory plant communities of the boreal mixedwood forest 2 and 34 years after disturbance. Canadian Journal of Forest Research 37, 2637–2651.

Kuuluvainen, T. (2002) Natural variability of forests as a reference for restoring and managing biological diversity in boreal Fennoscandia. Silva Fennica 36, 97–125.

Kuuluvainen, T., Aapala, K., Ahlroth, P. et al. (2002). Principles of ecological restoration of boreal forested ecosystems: Finland as an example. Silva Fennica, 36, 409–422.

Kuuluvainen, T. and Grenfell, R. (2012) Natural disturbance emulation in boreal forest ecosystem management – theories, strategies, and a comparison with conventional even-aged management. Canadian Journal of Forest Research 42, 1185–1203.

Landres, P.B., Morgan, P., and Swanson, F.J. (1999) Overview of the use of natural variability concepts in management ecological systems. Ecological Applications 9, 1179–1188.

Lindenmayer, D.B., Burton, P.J., and Franklin, J.F. (2008) Salvage Logging and its Ecological Consequences. Island Press, Washington, DC.

Lindenmayer, D.B., Foster, D.R., Franklin, J.F. *et al.* (2004) Salvage harvesting policies after natural disturbance. Science 303, 1303.

Lindenmayer, D.B. and Noss, R.F. (2006) Salvage logging, ecosystem processes, and biodiversity conservation. Conservation Biology 20, 949–958.

Linder, P., Jonsson, P. and Niklasson, M. (1998) Tree mortality after prescribed burning in an old-growth Scots pine forest in northern Sweden. Silva Fennica, 32, 339–349.

Liu, Z., He, H.S., and Yang, J. (2012) Emulating natural fire effects using harvesting in an eastern boreal forest landscape of northeast China. Journal of Vegetation Science 23, 782–795.

McCullough, D.G., Werner, R.A., and Neumann, D. (1998) Fire and insects in northern and boreal forest ecosystems of North America. Annual Review of Entomology 43, 107–127.

McIver, J.D. and Starr, L. (2001) A literature review on the environmental effects of postfire logging. Western Journal of Applied Forestry 16, 159–168.

Minshall, G.W. (2003) Responses of stream benthic macroinvertebrates to fire. Forest Ecology and Management 178, 155–161.

Moore, R.D. and Richardson, J.S. (2012) Natural disturbance and forest management in riparian zones: comparison of effects at reach, catchment, and landscape scales. Freshwater Science 31, 239–247.

Morissette, J.L., Cobb, T.P., Brigham, R.M., and James, P.C. (2002) The response of boreal forest songbird communities to fire and post-fire harvesting. Canadian Journal of Forest Research 32, 2169–2183.

Murphy, E.C. and Lehnhausen, W.A. (1998) Density and foraging ecology of woodpeckers following a stand-replacement fire. Journal of Wildlife Management 62, 1359–1372.

Nappi, A., Déry, S., Bujold, F. *et al.* (2011) Harvesting in burned forests – Issues and orientations for ecosystem-based management. Québec, Ministère des Ressources naturelles et de la Faune, Direction de l'environnement et de la protection des forêts, Québec, QC.

Nappi, A. and Drapeau, P. (2009) Reproductive success of the black-backed woodpecker (*Picoides arcticus*) in burned boreal forests: are burns source habitats? Biological Conservation 142, 1381–1391.

Nappi, A. and Drapeau, P. (2011) Pre-fire forest conditions and fire severity as determinants of the quality of burned forests for deadwood-dependent species: the case of the black-backed woodpecker. Canadian Journal of Forest Research 41, 994–1003.

Nappi, A., Drapeau, P., and Savard, J.P.L. (2004) Salvage logging after wildfire in the boreal forest: is it becoming a hot issue for wildlife? Forestry Chronicle 80, 67–74.

Niemelä, J. (1999) Management in relation to disturbance in the boreal forest. Forest Ecology and Management 115, 127–134.

Noss, R.F. (1987) From plant communities to landscapes in conservation inventories: a look at the Nature Conservancy (USA). Biological Conservation 41, 11–37.

Omnibus Public Land Management Act (2009) Title IV (Pub. L. 111–11, H.R. 146).

OMNR (2001) Forest management guide for natural disturbance pattern emulation, Version 3.1. Ontario Ministry of Natural Resources, Forest Management Branch, Toronto, ON.

OMNR (2010) Forest management guide for conserving biodiversity at the stand and site scales. Ontario Ministry of Natural Resources, Forest Management Branch, Toronto, ON.

Parker, T.J., Clancy, K.M., and Mathiasen, R.L. (2006) Interactions among fire, insects and pathogens in coniferous forests of the interior western United States and Canada. Agricultural and Forest Entomology 8, 167–189.

Parviainen, J. (1996) The impact of fire on Finnish forests in the past and today. In Goldammer, J.G. and Furyaev, V.V. (eds) Fire in Ecosystems of Boreal Eurasia. Kluwer Academic Publishers, Dordrecht, Forestry Sciences Vol. 48. pp. 55–64.

Perera, A.H. and Buse, L.J. (2004) Emulating natural disturbance in forest management: An overview. In Perera, A.H., Buse, L.J. and Weber, M.G. (eds) Emulating Natural Forest Landscape Disturbances: Concepts and Applications. Columbia University Press, New York, NY. pp. 3–7.

Perera, A.H., Buse, L.J., and Weber, M.G. (eds) (2004a) Emulating Natural Forest Landscape Disturbances: Concepts and Applications. Columbia University Press, New York, NY.

Perera, A.H., Buse, L.J., Weber, M.G., and Crow, T.R. (2004b) Emulating natural forest landscape disturbances: a synthesis. In Perera, A.H., Buse, L.J., and Weber, M.G. (eds) Emulating Natural Forest Landscape Disturbances: Concepts and Applications. Columbia University Press, New York, NY. pp. 265–274.

Perera, A.H. and Cui, W. (2010) Emulating natural disturbances as a forest management goal: lessons from fire regime simulations. Forest Ecology and Management 259, 1328–1337.

Peterson, D.L., Agree, J.K., and Aplet, G.H. (2009) Effects of timber harvest following wildfire in western North America. USDA Forest Service, Pacific Northwest Research Station, Portland, OR. General Technical Report PNW-GTR-776.

Phillips, I.D., Cobb, T.P., Spence, J.R., and Brigham, R.M. (2006) Salvage logging, edge effects, and carabid beetles: connections to conservation and sustainable forest management. Environmental Entomology 35, 950–957.

Pilliod, D.S., Bury, R.B., Hyde, E.J., Pearl, C.A., and Corn, P.S. (2003) Fire and amphibians in North America. Forest Ecology and Management 178, 163–181.

Purdon, M., Brais, S. and Bergeron, Y. (2004) Initial response of understorey vegetation to fire severity and salvage-logging in the southern boreal forest of Québec. Applied Vegetation Science 7, 49–60.

Pyne, S.J. (1996) Wild hearth: A prolegomenon to the cultural fire history of northern Eurasia. In Goldammer, J.G. and Furyaev, V.V. (eds) Fire in Ecosystems of Boreal Eurasia. Kluwer Academic Publishers, Dordrecht, Forestry Sciences Vol. 48. pp. 21–44.

Reeves, G.H., Bisson, P.A., Rieman, B.E., and Benda, L.E. (2006) Postfire logging in riparian areas. Conservation Biology 20, 994–1004.

Saab, V.A., Russell, R.E., and Dudley, J.G. (2009) Nest-site selection by cavity-nesting birds in relation to postfire salvage logging. Forest Ecology and Management 257, 151–159.

ROLE OF WILDFIRE RESIDUALS IN FOREST MANAGEMENT APPLICATIONS

Saab, V.A., Russell, R.E., Rotella, J.J., and Dudley, J. (2011) Modeling nest survival of cavity-nesting birds in relation to post-fire salvage logging. Journal of Wildlife Management 75, 794–804.

Saint-Germain, M. and Greene, D.F. (2009) Salvage logging in the boreal and cordilleran forests of Canada: integrating industrial and ecological concerns in management plans. Forestry Chronicle 85, 120–134.

Schmiegelow, F.K.A., Stepnisky, D.P., Stambaugh, C.A., and Koivula, M. (2006) Reconciling salvage logging of boreal forests with a natural-disturbance management model. Conservation Biology 20, 971–983.

Sibley, P.K., Kreutzweiser, D.P., Naylor, B.J., Richardson, J.S., and Gordon, A.M (2012) Emulation of natural disturbance (END) for riparian forest management: synthesis and recommendations. Freshwater Science 31, 258–264.

Silins, U., Stone, M., Emelko, M.B., and Bladon, K.D. (2009) Sediment production following severe wildfire and post-fire salvage logging in the Rocky Mountain headwaters of the Oldman River Basin, Alberta. Catena 79, 189–197.

Similä, M. and Junninen, K. (2012) Ecological restoration and management in boreal forests – best practices from Finland. Metshallitus Natural Heritage Services, Helsinki, Finland. Available from http://julkaisut.metsa.fi/assets/pdf/lp/Muut/ecological-restoration.pdf (accessed August 2012).

Toivanen, T. and Kotiaho, J.S. (2007) Mimicking natural disturbances of boreal forests: the effects of controlled burning and creating dead wood on beetle diversity. Biodiversity and Conservation 16, 3193–3211.

Toivanen, T. and Kotiaho, J.S. (2010) The preferences of saproxylic beetle species for different dead wood types created in forest restoration treatments. Canadian Journal of Forest Research 40, 445–464.

Vanha-Majamaa, I., Lilja, S., Ryoma, R. et al. (2007) Rehabilitating boreal forest structure and species composition in Finland through logging, dead wood creation and fire: the EVO experiment. Forest Ecology and Management 250, 77–88.

Van Wilgenburg, S.L., and Hobson, K.A. (2008) Landscape-scale disturbance and boreal forest birds: can large single-pass harvest approximate fires? Forest Ecology and Management 256, 136–146.

Virkkala, R., Alanko, T., Laine, T., and Tiainen, J. (1993) Population contraction of the white-backed woodpecker (Dendrocopos leucotos) in Finland as a consequence of habitat alteration. Biological Conservation 66, 47–53.

Wallenius, T. (2011) Major decline in fires in coniferous forests – reconstructing the phenomenon and seeking for the cause. Silva Fennica 45, 139–155.

Wikars, L.-O. (2002) Dependence on fire in wood-living insects: an experiment with burned and unburned spruce and birch logs. Journal of Insect Conservation 6, 1–12.

6 Ecology of boreal wildfire residuals – a summary and synthesis

Our goal in this concluding chapter is to briefly summarize what is known about the types and ecology of tree-level wildfire residuals in the boreal forest and how this knowledge can inform forest management, without repeating the specific details presented in previous chapters. We also explore what is not known; that is, we identify some of the concepts and questions related to boreal wildfire residuals that have not received much attention from researchers.

Wildfire residuals and their occurrence

The wildfire disturbance regime is an inherent characteristic of the boreal forest biome. It is an abiotic process that typically occurs over extensive areas, is both

Ecology of Wildfire Residuals in Boreal Forests, First Edition. Ajith H. Perera and Lisa J. Buse.
© 2014 Ajith H. Perera and Lisa J. Buse. Published 2014 by John Wiley & Sons, Ltd.
Companion Website: www.wiley.com/go/perera/wildfire

spatially and temporally frequent, and leads to secondary forest ecosystem succession. Boreal forest ecosystems have evolved within this disturbance regime and, as a result of this regime, are self-organized at broader scales in a continuously shifting spatial mosaic. Residual vegetation resulting from wildfires is a key component of ecosystem revival in the short term and the evolutionary process in the long term. The structure and ecology of these wildfire residuals are complex and highly variable. No two boreal fire footprints are alike, since the fire process is highly variable and, therefore, the residuals are an emergent property.

Residual vegetation types

In this book, we have included all intact vegetative parts that exist after a wildfire as residuals. This view is broader than descriptions of wildfire residuals common in literature, in which residuals have been considered narrowly, as studies of residuals were typically motivated by a specific ecological process. Furthermore, the terms that have been used in the literature not only vary widely, but are inconsistent in definition, so that they do not provide a ready basis for among-study comparisons and contrasts. Also, use of the term "fire residual" in the literature appears to be implicitly reserved for tree vegetation.

There is clearly a need for a more robust and explicit definition of wildfire residuals and one that is broadly applicable across scales. Ideally, such a conceptual framework should be based on a logic that goes beyond purely descriptive characteristics of the residual vegetation. Therefore, we have based our definitions of wildfire residuals on the common foundation of how they were formed, and categorized them by a consistently measurable variable – the degree of fire severity (Figure 6.1). Even though our primary focus has been on tree-level residuals, our proposed approach to defining, identifying, and measuring wildfire residuals can be extended to understory and forest floor vegetation.

Residual patches

Residual vegetation patches are intact remnants of the pre-burn ecosystem that occur within wildfire footprints only because they have escaped the fire. These areas show no signs of fire, and occur as unburned islands within the matrix of the burned area of a wildfire footprint in an array of sizes and shapes that result from variations in the spatial patterns of spread and extinguishment of the local fire front. Such escapees occur at a range of spatial scales. Local fire fronts may extinguish for a few meters to hundreds of meters, producing residual patches ranging in size from a few square meters of forest floor to hundreds of hectares of intact forest. Although we found no formal studies of fine-scale residual patches

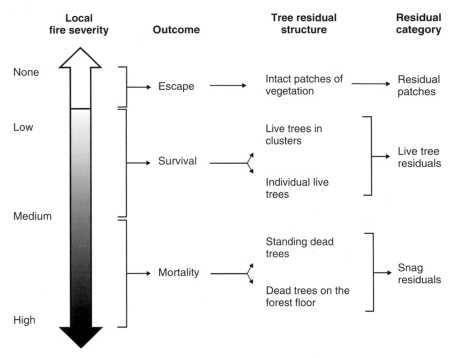

Figure 6.1   Boreal wildfire residuals can be categorized based on the local fire severity and their spatial arrangement. This approach offers a consistent basis to define, study, and assess tree-level residuals, as shown here, but can also be applied to other residual vegetation at finer scales.

in the literature, surveys of tree-based residual patches are abundant. The extents of the residual patches in these reports range from none to nearly a quarter of the area of the wildfire footprint.

Such results, however, must be compared and generalized with caution, because the definitions and mapping methods employed vary considerably among these surveys. Even so, it is certain that almost all boreal wildfire footprints contain residual patches, except perhaps under extreme fire weather scenarios. The composition of the residual patches varies, and may not represent an exact replica of all pre-burn communities within the overall wildfire footprint. This is because residual patches may occur with spatial biases toward topographical features such as a high water table or other barriers to fire spread, where fire fronts are likely to be locally diverted or extinguished, especially during mild fire weather. Once residual patches have formed during the fire, unless burned by a subsequent fire event, they stay intact for many decades and have been observed to remain distinct from the surrounding forest even 75 years later. Other than natural succession of the species composition, the only changes that may occur in these residual patches are windthrow and some degree of delayed mortality of live trees at their perimeters.

Live tree residuals

Some individual trees survive exposure to fire due to various adaptive morphological and physiological mechanisms that provide fire tolerance. Characteristically, these trees have a partially or entirely green crown, and show minor or few signs of scorching. Live tree residuals are likely to occur mostly in areas of low or medium fire severity, where the fire intensities were too low to overcome the tolerance mechanisms of trees and cause lethal damage. While biases to thick-barked species and older ages have been commonly reported with respect to live tree residual composition, it is probable that such reports are based on circumstantial evidence and may not take into account the context of local fire intensity. Live tree residuals may occur either as individual stems or as clusters, and can be found scattered throughout a wildfire footprint, but are especially prevalent in transition zones between burned and unburned areas.

It is not uncommon for clusters of live trees to be confounded with residual patches in the literature. Although they may appear structurally similar from a distance, we consider such live tree clusters in burned areas with low fire severity to be distinct from residual patches that are unburned areas because their origin and post-fire ecological functions differ. Estimates of the abundance of live tree residuals in the literature vary, ranging from 0 to 700 stems per hectare, although the specific definitions of what constitutes a live tree are not consistent among studies. A notable characteristic of live tree residuals is that they are the most transient of all wildfire residuals. They undergo delayed mortality and windthrow in the first few years after a fire; during this period the population of live tree residuals may decrease by as much as 60% from delayed mortality and 15% from windthrow. Therefore, reported estimates of the abundance of live tree residuals must be considered in relation to the time of their survey after the fire.

Snag residuals

Snag residuals are the trees that die from exposure to the fire, yet are not completely incinerated. They show no signs of life, and may or may not remain rooted to the ground. Their fine branches may be consumed by the fire, but their large branches are usually retained. Snag residuals that are uprooted or broken and that lie on the forest floor are broadly termed "downed wood". Though the majority of snag residuals form during the fire, their formation continues as live tree residuals die, especially during the first few years after the fire. This delayed recruitment of snags from live tree residuals is likely to depend on the initial fire severity, as well as the species and age of the tree.

Snags are the most abundant of all boreal wildfire residuals, and occur throughout the wildfire footprint wherever the fire severity was medium or high. Estimates of snag abundance are common in the literature, but the metrics and stem dimensions used to define a snag are highly inconsistent. Furthermore, species tallies and other characteristics commonly detailed in the literature provide insufficient information because they do not take into account the contextual pre-burn forest community characteristics, which are usually not reported.

General observations

Overall, it is difficult to compare the many reports of boreal wildfire residuals in the scientific literature and reach general conclusions. Impediments to comparison include inconsistencies in defining wildfire residual categories and in the metrics used to quantify their abundance, as well as differences in how long after the fire the surveys were conducted. Moreover, the published reports often contain no information on the pre-burn ecosystem or on fire severity patterns in detail. Still, there are some common observations about boreal wildfire residuals:

- All wildfire footprints contain snag residuals in areas of medium to high fire severity and residual patches where fire fronts have changed direction or were locally extinguished. Live tree residuals occur in areas of low to medium fire severity, such as transition zones between snag residuals and residual patches.
- The relative abundance of residual components is highly variable, and is based on a multitude of factors that operate at different scales to influence fire behavior and fire severity.
- Wildfire residuals, once they have formed during the fire, are not static. They continue to change over time after the fire. Live tree residuals are the most dynamic and transient component, as they rapidly die to become snags. Furthermore, live tree and snag residuals gradually fall to become downed wood, and decompose over a long period. Residual patches are the most stable of the residual categories, except for changes at their perimeters due to windthrow.

Spatial patterns of residuals

The formation of residuals during a wildfire is the outcome of a complex interaction between the vegetation and the intensity of the fire front. Many biotic and abiotic variables influence this interaction. These include vegetation characteristics such as species, age, and stand structure; local geo-environmental factors such as the water table, slope, and aspect; and weather factors both before and during the fire. The weather comprises elements such as drought, wind, precipitation, and lightning that originate and influence the fire event from beyond the scale of the fire footprint, as well as elements that originate within the fire footprint, most notably intra-fire whirlwinds. Not only do these variables operate at different spatial and temporal scales, they also interact both factorially (at the same scale) and hierarchically (nested at different scales).

It is likely that macro-weather, which comprises a region's climate and seasonality, will encompass meso-weather, which includes local rainfall, wind, and temperature, and will in turn encompass micro-weather, which is composed of factors such as within-fire winds. The local geo-environment and the vegetation characteristics may influence the formation of residuals only in the context of these nested weather factors. Moreover, their interactions are not linear, and may involve many thresholds. For example, when the fire weather is extreme, the influence of these factors on the probability of the occurrence of areas that escape fire may diminish non-linearly (Figure 6.2). These thresholds may be associated with different steps in the fire's behavior (e.g., ignition, combustion, crowning, spotting) as well as the vegetation response to the fire (e.g., the lethal temperature and fire residence time). Boreal wildfire residuals are therefore an emergent property of the fire, and are not necessarily predictable from all or even the sum of the known variables.

Awareness that residual formation is a complex process, and that the spatial patterns of wildfire residuals are not readily predictable from pre-burn vegetation and geo-environment, is not commonly evident in the scientific literature. Many studies, based on forensic analyses of fire footprints, include claims that vegetation characteristics and geo-environmental features have caused the abundance and spatial patterns of wildfire residuals. These studies, combined with observations from only one or a few fire footprints, have led to certain popular beliefs. These include the belief that residual patches always occur on the leeward side of fuel barriers, that they have a spatial bias toward wet areas and edges of water bodies, or that certain species will produce more live tree residuals. Even if such biases have been observed in some cases, possibly based on spurious spatial correlations, it is unlikely that the processes that led to such residual patterns will be repeated in another fire event. That is because the spatial configuration of vegetation and the geo-environment, as well as the macro-, meso-, and micro-scale weather, will inevitably differ.

We believe that it is futile to search for single determinants of the abundance and spatial distribution of wildfire residuals, factors that operate at a particular

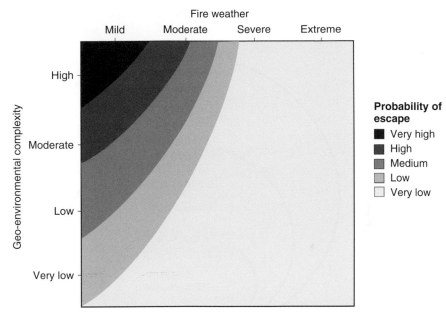

Figure 6.2   A hypothetical depiction of the probability of occurrence of areas that escape fire within a boreal wildfire footprint, based on the non-linearly interactive influence of environmental complexity and the severity of the fire weather. When fire weather is mild, high site complexity helps more areas to escape the fire by presenting local fire barriers, but that effect is overridden when fire weather becomes more extreme.

scale. Closely related to this pursuit of a simple explanation of abundance and distribution is a common mistake made by those who use wildfire residual patterns as a model for designing forest harvest patterns: the mistake of prescribing "normal" residual abundances and spatial biases. Ideally, an algorithm should be used to generate the spatial probability of the occurrence of residuals within a fire footprint, an algorithm based on vegetation, geo-environment, and likelihood of macro-, and meso-scale fire weather. Even with use of an algorithm, however, many factors that determine the formation of residuals may not be predictable with any degree of certainty. Such factors include various properties of fire behavior, such as spotting, that emerge during the fire event, and micro-scale weather such as whirlwinds.

The concept of wildfire residuals, moreover, should not be limited to the scale of trees or the fire footprint. Many residuals exist at scales finer than trees. For example, the residual structures of some shrubs, ground flora, and even fungi, remain unincinerated and intact, either on the forest floor or below the ground,

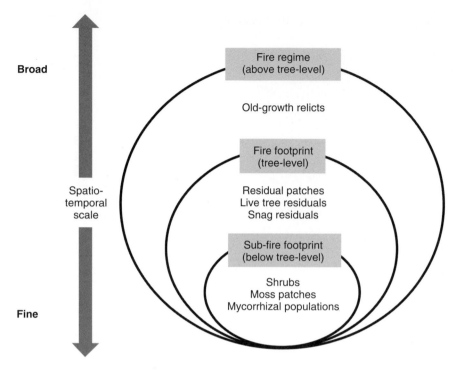

Figure 6.3   Boreal wildfire residuals exist beyond the scale of fire footprints and trees, even though residuals that form at both finer and broader scales have not attracted the attention of researchers.

and these are likely to play important roles in ecosystem recovery in the short term and to contribute to biodiversity in the long term. Yet, these residuals have eluded the attention of researchers, as is evident from the paucity of related scientific literature.

It is also true that wildfire residuals exist at spatial and temporal scales much broader than the single fire footprint (Figure 6.3). Large areas of boreal forest may continue to escape wildfires because of spatial randomness in fire ignition and spread, or spatial biases in the weather and geo-environmental factors. Regardless of the cause, such escapees may exist for long periods at fire-regime scale, that is, at the scale of patterns created by multiple fire events over large extents and long periods of time – and these forest areas commonly become "old-growth forest". Such areas may repeatedly escape fire, at least until some key aspect of the context changes, such as the climate or the ignition source. Although the ecological value of these old-growth relicts in promoting heterogeneity and conserving rare habitats and species at the broader landscape scale are well documented, they have not been considered in the scientific literature in the context of wildfire residuals.

ECOLOGY OF BOREAL WILDFIRE RESIDUALS – A SUMMARY AND SYNTHESIS

Ecological roles of wildfire residuals

Life returns to a fire footprint while parts of it are still smoking, and the recovery of the burned ecosystem begins immediately after the fire. It is evident that wildfire residuals initiate and continue to play an important role in most, if not all, of the processes involved in this recovery. Some processes occur only locally, while others extend even beyond the fire footprint to the regional scale. Nonetheless, these ecological processes may be linked to a specific category of residuals. The most prominent residual category in a boreal wildfire footprint is snag residuals, followed by residual patches. While live tree residuals do occur, they constitute a very small fraction of the wildfire footprint and very rapidly transition to snags. As such, their ecological roles are less clear, and perhaps less significant than those of snag residuals and residual patches.

Snag residuals

Perhaps the most visible of all ecological processes associated with snag residuals is their hosting of saproxylic beetles and the woodpeckers that feed on the beetles. After the woodpeckers depart, the snag residuals also host secondary cavity-nesting species. And although serotinous and semi-serotinous conifer residuals act as a vast *in situ* aerial seedbank to assist regeneration of burned sites, some deciduous species do so via stem, collar, or root sprouts. Snag residuals further aid the recovery of plant communities by slowly releasing nutrients stored in their stems and by ameliorating the microclimate. As well, these dead stems and their roots represent a vast reservoir of carbon that buffers the rapid emission of carbon dioxide during a fire event. Although these examples may broadly capture the ecological roles associated with snag residuals, many other specific ecological processes are likely yet to be discovered, especially those associated with other forms of vegetation such as mosses and shrubs.

Though the exact timing of the onset and end of these specific processes is not certain, the duration of the ecological roles of snag residuals is known to vary (Figure 6.4). Some roles commence immediately after the burn, such as the hosting of pyrophilous beetles, but do not last beyond a few years. Other roles, such as the provision of propagules for regeneration, also begin immediately after the fire, but continue for years. Still others, such as slow release of nutrients, commence relatively late and continue for decades. Ecological processes associated with snag residuals are not limited to the locales where they occur within wildfire footprints, but also play a role at a regional level; for example, in support of the continuity of the population dynamics of saproxylic beetles and woodpeckers. Therefore, disruptions of snag residuals, such as their removal by salvage logging, may have regional consequences in addition to the obvious local impacts.

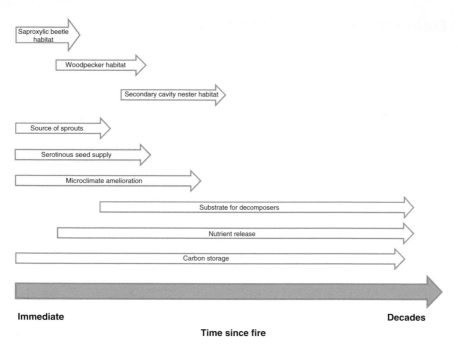

Immediate   **Decades**

Time since fire

Figure 6.4   Ecological roles of snag residuals commence immediately after a fire and some continue for decades, until the residual stems are completely decomposed.

Residual patches

A widely perceived role of residual patches is their buffering effect on surface runoff after fires and their mitigation of the subsequent sedimentation of lakes and streams. In addition, it has been hypothesized that residual patches at the edges of bodies of water minimize the heating and turbulence effects of exposure on aquatic biota. Moreover, this role may extend to minimizing the impacts of fire on streams at a watershed scale, well beyond a fire footprint. Residual patches also shelter mobile animals during the fire event, and provide them with temporary habitats for some period following the fire. Because the residual patches are exact replicas of local biotic communities, they contain pre-burn assemblages of plants and animals, and allow many ecological processes that occurred prior to the fire event to continue after the fire. This long-term role of residual patches contributes to the recovery of the burned ecosystem as a source of biota – both plant propagules and animals – for its repopulation. As well, it is thought that residual patches act as refuges for many species until the burned system recovers. This "lifeboating" has a broader significance in maintaining

ECOLOGY OF BOREAL WILDFIRE RESIDUALS – A SUMMARY AND SYNTHESIS

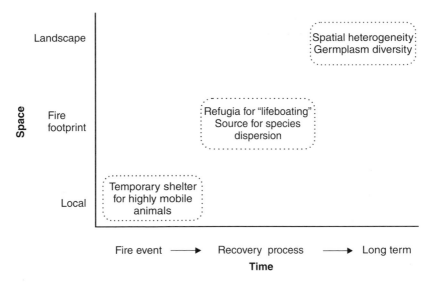

Figure 6.5   The roles that residual patches may play in the recovery of the post-fire ecosystem vary both spatially (from local to landscape-level effects) and temporally (over periods potentially ranging from years to centuries).

species assemblages that would otherwise be eradicated by fire, especially if these events are frequent. Consequently, residual patches may act as centers of species diversity that contribute to biodiversity patterns at larger spatial scales. Residual patches that escape repeated fires may even have evolutionary significance due to their harboring of germplasm for centuries.

Interspersion of residual patches within a wildfire footprint forms a network of unburned areas. These dispersed islands may decrease the relative distance for animal movement across burned areas, and increase the opportunity for windborne seed dispersal from unburned areas into the burned area. Much remains unknown about this "stepping stone" role for ecological flows, but it may conceivably continue until the burned system is fully recovered. Given that residual patches may last for centuries, their roles obviously continue for a long period after a fire event. This would be especially true at broader scales (Figure 6.5).

A conceptual view

In an attempt to summarize the various ecological processes that may involve boreal wildfire residuals, we have grouped them into enhancements and reductions of ecological processes, cyclical processes, and flows of biotic and

ECOLOGY OF BOREAL WILDFIRE RESIDUALS – A SUMMARY AND SYNTHESIS

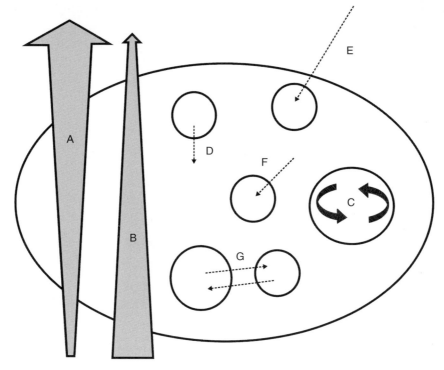

Figure 6.6   An abstract depiction of the various ecological roles, both well-known and hypothetical, of boreal wildfire residuals (circles) in relation to the burned area within the wildfire footprint (the large oval) and the unburned area outside the footprint. These include: (A) enhancements and (B) reductions of ecological processes in the context of the burned and unburned matrix within the broader landscape; (C) cycles within the residuals; (D) flows from residuals into the burned area; (E) flows toward residuals from unburned area outside the wildfire footprint; (F) flows toward residuals from the burned area; and (G) flows among residuals.

abiotic materials (Figure 6.6). An example of an *enhancement* is the multiplication of the pyrophilous insects and woodpeckers that use fires, which may arrive, multiply, and depart in greater numbers. Slowing erosion and surface runoff and decreasing sedimentation of streams and lakes due to buffering by residuals is an example of a *reduction*. Maintaining ecological processes from the pre-burn ecosystem, such as within residual patches, is an example of a *cycle*. In addition, there are four types of *flows*. Some materials flow from residuals into the burned area, as in the case of dispersing propagules of mycorrhizae, lichen, and trees. Others flow from outside the fire footprint to the residuals, as in the example of the arrival of ungulates to browse the newly sprouting vegetation and cavity nesters to occupy nests of woodpeckers that have departed the fire footprint. Movement of highly mobile animals to residual patches during the burn is an example of flow towards residual patches from within the burn. Finally, flows occur among the residuals; for example, some bird species nest in residual patches but feed on the insects and rodents that live among the snag residuals.

Knowledge uncertainties

Only some of the ecological roles performed by wildfire residuals are well understood and supported by a substantial body of scientific evidence. However, ecologists commonly expect residuals to play many ecological roles beyond the well-established, such as supplying propagules for regeneration and initiating the sequence from saproxylic beetles to woodpeckers and secondary cavity nesters. When the state of knowledge of these ecological roles is scrutinized from the literature, however, it is apparent that few of these expectations have a solid foundation in scientific evidence; thus, they remain mostly assumptions and untested hypotheses. The role of residual patches in mitigating the impacts of exposure on bodies of water is a good example of a common expectation that does not seem to have ample and direct scientific evidence.

In particular, many knowledge gaps exist with respect to the broader ecological roles of residuals in the context of fire-driven boreal forest landscapes. These include the concept of "lifeboating", in which residual patches harbor and shelter organisms – both plants and animals – until they can repopulate the burned area once it has partly or fully recovered. Another such role is the perpetuation of organisms and biotic sequences that depend on wildfire residuals in a regional context. It is conceivable that populations of these organisms shift spatially within a fire-driven region, as long as there is an adequate supply of wildfire residuals at favorable extents and intervals. The carbon dynamics of all categories of residuals must be important contributors to regional and global carbon fluxes. However, beyond gross estimates and assumptions little is known about carbon stocks in wildfire residuals and their fate. Residual patches produce a spatial mosaic within wildfire footprints that must influence many spatial flows and the movement of organisms. For example, residual patches must provide stepping stones that support animal movement from outside the footprint to the internal parts of a burned area, and aid the repopulation process. However, such topics remain unexplored to date. Also, the consequences of the enhanced landscape heterogeneity at a meso-scale in terms of biodiversity, system resilience, and evolution are presently unknown.

Management applications and wildfire residuals

Salvage logging

Salvage logging is a centuries-old practice in which burned and unburned trees are extracted from wildfire footprints. Because of increased access to burned areas, this practice has become so common recently that it has drawn much attention from scientists, who have rallied globally during the last decade to question the ecological validity of this practice. Their primary concerns are that wildfire residuals are critically important to the recovery of the burned ecosystem on a local scale, and to the ecological chain of pyrophilous beetles, woodpeckers, and cavity nesters on a regional scale, as these organisms rely on a continuous supply of intact burned sites. Although there is a body of literature that highlights the detrimental aspects of post-fire harvesting of residuals and includes pleas to discontinue this practice, no published research explicitly endorses the practice, either from an ecological or an economic perspective. Still some continue to view wildfire residuals as an ecosystem that has been destroyed and that is, therefore, of little ecological value.

This state of imbalance is exacerbated by the many uncertainties in the scientific knowledge, which do not inspire a high degree of confidence about either the impacts or merits of salvage logging. Furthermore, due to the high variability inherent in wildfires, post-fire conditions, and logging practices, it is difficult to design and conduct studies to test specific hypotheses about the consequences of salvage logging and to generalize these findings, and it would be many years before such studies could yield results. Moreover, some ecologists have alleged that many government agencies continue tacitly to promote salvage logging, or to provide only ambiguous direction about it, despite the lack of evidence that this activity is ecologically innocuous. Forest managers who harvest wildfire residuals appear to do so with little assistance from scientific knowledge and with weak guidance from policymakers about the timing, type, and locations of the salvage harvest. Consequently, salvage logging after wildfires proceeds for economic reasons, even though claims about the ecological impacts of this practice remain untested and, as a result, the controversy that surrounds the harvesting of wildfire residuals continues.

Emulating wildfire disturbances

Using fire disturbance patterns as a template to design forest harvest plans is a recent trend in the management of fire-driven landscapes in boreal North

America and Fennoscandia. Patterns gleaned from wildfire residuals provide the basis for this practice at the within-harvest scale. The extents and spatial patterns of residual patches inform the amounts, sizes, shapes, and locations of the patches of forest that will be left unharvested, and the abundance and interspersion of live tree residuals determine which individual trees are not felled. Foresters expect that the standing dead stems and downed wood that result from harvesting will emulate some portion of the patterns of the snag residuals in wildfires. The premise of this forest management paradigm is that, by emulating wildfire residual patterns, the footprint of the forest harvest will be similar to that of a wildfire, and the ecological processes that occur will resemble those in wildfire footprints. In the absence of perfect ecological knowledge, this approach is expected to be precautionary, and it offers a coarse-filter solution to conserving biodiversity. Even though this argument is conceptually sound, the practice of emulating wildfire disturbance encounters many challenges.

Foremost among these challenges is the lack of direct scientific evidence that emulating wildfire disturbance succeeds in conserving biodiversity; similarly, proof is lacking that the practice of leaving residual structure in a harvest area accomplishes the ecological goal of inducing the same "natural" ecological processes that follow wildfire. There are also numerous practical difficulties involved in emulating wildfire residual patterns. First, there is uncertainty about how much and where wildfire residuals occur; secondly, the tactical realities of reproducing such spatial patterns of residuals within a forest harvest pose a daunting challenge. In practice, emulation goals must be based on a range of values and probabilities; that is, emulating variability is more important than retaining average values of residuals. However, forest managers appear to prefer simple and deterministic patterns instead of complex and probabilistic ones. These limitations, coupled with conflicts with other forest management goals, differing expectations by environmentalists, and the wide range of societal attitudes, present many obstacles to emulating wildfire disturbance.

Although retaining and emulating the wildfire residual structure during harvest is perhaps better than engineering forest harvest with a single focus, such as economic values or a desired species, the major question appears to be whether leaving residual structure within the footprints of forest harvests truly enhances the heterogeneity of the footprint and catalyzes key ecological processes. Improved ecological knowledge of wildfire residuals – both patterns and processes – will help to answer this question, and will provide a sound basis for the practice of emulating wildfire disturbances. Furthermore, incorporating an adaptive management approach by monitoring outcomes and revising practices would iteratively improve strategies to emulate wildfire.

ECOLOGY OF BOREAL WILDFIRE RESIDUALS – A SUMMARY AND SYNTHESIS

Restoring wildfire residuals

In areas of the boreal forest where wildfire occurrence has been nearly eliminated by human activities such as fire suppression, concern among ecologists about the associated loss of biodiversity has generated interest in the forest management strategy of reintroducing fire to the landscape. The expectation is that the use of fire will restore key ecological processes and bolster fire-dependent species. In Fennoscandia, for example, reintroducing fire is seen as a way to recreate specific forest habitat types that result from wildfires. The main focus there has been creating residual snags and downed wood to supply habitat for pyrophilous beetles, which have been deemed critically important for conservation after many decades of aggressive fire suppression.

Despite some local successes for some species and sites, the challenges of reintroducing fire are numerous. The most problematic is that of scale. Even though wildfire disturbance is a landscape-level process, much of the direction for and implementation of the restoration of fire has, for practical reasons, been focused at the scale of the individual fire event. As a result, even if some local processes are successfully restored through the use of fire, regional-level processes may not be, and thus restoration efforts may not fully achieve the intended results.

In practice, the reintroduction of fire to restore wildfire residuals is hampered by incomplete knowledge of the effects of boreal wildfires on ecosystems. Wildfires are not only probabilistic but also highly heterogeneous, making it difficult to plan when, where, and how much restoration burning is appropriate. Reproducing the range and variability of conditions resulting from wildfire disturbance is not feasible with managed fires, so choices are necessary about what aspects or characteristics of wildfire to restore. Also, as we discussed previously, the patterns and degrees of fire severity within a fire footprint are highly variable and depend on a multitude of interacting factors. Given that the timing, intensity, and duration of managed fires will differ from those of wildfires, the resulting ecological effects may also differ; therefore, the desired residual structure and ensuing ecological processes may not be restored.

Beyond the knowledge uncertainties, reintroducing fire requires changes in forest management policies, which are often difficult to achieve and which depend to some extent on societal acceptance of the proposed practices. Given that fire is generally viewed as a destructive phenomenon and something to be avoided, especially near human settlements, policies that advocate reintroducing fire to the landscape may meet with strong public opposition. The knowledge uncertainties, combined with the policy, societal, and logistical challenges of reintroducing fire to forest landscapes, means that this management approach is likely to have limited application, except as the last resort for conserving fire-dependent species that rely on wildfire residuals.

Suppressing fire

Eliminating wildfires has serious impacts on species and ecological processes that depend on wildfire residuals, as shown from the experience in Fennoscandia. Despite the known detrimental effects of this near-elimination of wildfires from more than a century of aggressive fire suppression, many regions of boreal North America are now attempting similarly strong wildfire suppression practices in response to increasing social and economic pressures.

Hitherto, no attempts have been made to evaluate the ecological consequences of modifying the spatial and temporal aspects of boreal wildfire occurrence through fire suppression and the resultant effects on fire severity patterns within wildfires, the residual structure, and their associated ecological processes. This question is pertinent not only at the scale of an individual fire, but also at broader scales where the metapopulation dynamics of fire-dependent species are likely to become significant. Overall, the effects of fire-suppression policies on wildfire residuals and the ensuing ecological processes remain unexplored.

Research needs on wildfire residuals

Advancing research methods

Many ecologists, entomologists and ornithologists in particular, have contributed much to the present body of knowledge of boreal wildfire residuals, with respect to both their occurrence and their associated ecological processes. However, it is timely to consider how the approaches and methods used in studying the ecology of residuals can be improved.

First, researchers should exercise greater discipline in planning their studies, so that each new study contributes to a stepwise progression of knowledge and does not merely add to the existing accumulation of disconnected surveys and studies undertaken in isolation of past findings and syntheses of knowledge. In doing so, researchers will be able to explicitly address uncertainties and gaps in the knowledge and test hypotheses identified in previous studies.

Secondly, there is much room for improvement in the rigor of the research methods, beginning with an informed and objective definition of residuals. Formulating explicit hypotheses and testing them rigorously must be an integral part of all studies – going beyond enumerations of residual abundance and descriptions of patterns.

Thirdly, it is necessary to expand the sample space to capture variability across several fires rather than just one fire, or even part of one fire. Similarly, there is a need to go beyond reporting surveys conducted at a single point in time, and to begin monitoring trends over longer periods. Finally, it is imperative that the researchers' outlook be broadened to include contextual factors such as fire weather, fire behavior, and even unburned areas outside of fire footprints, in their investigations of wildfire residuals.

Expanding research topics

Although ecologists have continued to advance the knowledge of a few specific topics, many additional topics related to wildfire residuals may deserve investigation. These include both academic and applied research questions. There are many unknowns, for example, about the mechanisms and specific details of ecological processes such as residual formation, the occupation of residuals by many animal species, the release of nutrients and carbon by snag residuals, amelioration of the microclimate and prevention of site degradation due to sheltering by residuals, and the longevity of residuals. Although many logically valid hypotheses and assumptions exist about these processes, those hypotheses and assumptions need rigorous testing in carefully designed studies.

Perhaps the area where attention is most needed lies in the ecological processes related to boreal wildfire residuals that extend in scope beyond fire footprints and individual fire events. Given that wildfires and residuals are an integral part of boreal forest landscapes, it is essential to examine how the spatial patterns of residuals influence ecological flows across broader spatial extents that include the mosaic of fire footprints of various ages and their positions within the unburned landscape. In addition, the roles of residual patches in germplasm conservation and their contributions to genetic, species, and community diversity, especially in terms of evolutionary processes, must be further examined.

Many questions remain partly or completely answered with respect to forest management applications that involve residuals. The foremost need is for more certain knowledge about the ecological consequences of salvage logging. Significant topics that need urgent research attention include the impacts of removal of residual trees and patches at the scale of an individual fire, such as the effects on tree regeneration, soil erosion, and lifeboating, and the impacts at a landscape scale, such as the effects on the regional population dynamics of pyrophilous insect species and the breeding and dispersal success of woodpeckers. Some of these questions will require examination of the frequency and extent of fire footprints at regional scales, well beyond the study of residuals at their scale of formation and occurrence.

Conclusion

Our focus in this book was on the residual vegetation within intense wildfires, a disturbance that typically replaces forest communities, in the boreal biome with specific emphasis on tree-level residuals. This is not to say that ground fires that consume the understory vegetation with little damage to the overstory – a type of fire common in Eurasia and some parts of boreal North America – should be ignored. Even though those fires do not form residuals at an individual-tree level, they do form residuals in the understory and below the ground surface and deserve attention from ecologists.

Much of our discussion on the ecology of residuals was based on the information from boreal North America and Fennoscandia. This bias was due to the paucity of literature from other parts of the boreal biome, in particular from boreal Eurasia, which meant that we could not ascertain the state of knowledge and research needs beyond North America and Fennoscandia. We assume that, in time, as research information is globalized, a more comprehensive understanding of the ecology of wildfire residuals will lead to assessments across the boreal forest biome.

Furthermore, we expect that interest in boreal wildfire residuals and their ecology will expand. Even with a conservative estimate of 10×10^6 ha of boreal wildfires annually, with half of those being intense fires, at least 0.5×10^6 ha of residual patches and 4.5×10^6 ha of snag residuals are produced every year across the boreal biome. If, as many predict, boreal wildfires increase in extent and frequency in response to a changing climate, then the complex spatial patterns generated by wildfire residuals and their vast pool of biomass will influence ecological processes not only locally, but increasingly at landscape and regional scales.

A growing recognition of the importance of boreal wildfire residuals will likely prompt the answering of many questions on their ecology, especially at broader spatial and temporal scales. Specifically, we expect that the challenges presented to researchers by the complexity and variability of fire disturbances will be overcome, the hitherto narrow focus of research will be broadened, and the rigor of the research will improve. An essential step for this advancement is a coalescence and synthesis of the state of knowledge of boreal wildfire residuals. We hope this compilation will provide an impetus to commence that phase.

ECOLOGY OF BOREAL WILDFIRE RESIDUALS – A SUMMARY AND SYNTHESIS

Index

Note: page numbers followed by f or t refer to figures/photographs or tables

Ecology of Wildfire Residuals in Boreal Forests, First Edition. Ajith H. Perera and Lisa J. Buse.
© 2014 Ajith H. Perera and Lisa J. Buse. Published 2014 by John Wiley & Sons, Ltd.
Companion Website: www.wiley.com/go/perera/wildfire